ODOUR AND AMMONIA EMISSIONS FROM LIVESTOCK FARMING

Proceedings of a seminar held in Silsoe, United Kingdom,
26–28 March 1990

ODOUR AND AMMONIA EMISSIONS FROM LIVESTOCK FARMING

Edited by

V.C. NIELSEN

ADAS Mechanisation Unit, Silsoe, Bedford, UK

J.H. VOORBURG

IMAG, Wageningen, The Netherlands

and

P. L'HERMITE

Commission of the European Communities, Brussels, Belgium

Taylor & Francis
Taylor & Francis Group

LONDON AND NEW YORK

Published by Taylor & Francis
2 Park Square, Milton Park, Abingdon, Oxon OX14 4RN
52 Vanderbilt Avenue, New York, NY 10017

First issued in paperback 2020

Taylor & Francis is an imprint of Taylor & Francis

British Library Cataloguing in Publication Data

Odour and ammonia emissions from livestock farming.
I. Nielsen, V. C. II. Voorburg, J. H.
III. L'Hermite, P.
628.7466

ISBN 1-85166-717-2

Library of Congress CIP data applied for

Publication arrangements by Commission of the European Communities, Directorate-
General Telecommunications, Information Industries and Innovation, Scientific and
Technical Communication Unit, Luxembourg

EUR 13410

LEGAL NOTICE

Special regulations for readers in the USA

Publisher's Note

ISBN 13: 978-0-367-58008-7 (pbk)
ISBN 13: 978-1-85166-717-8 (hbk)

PREFACE

This was the final workshop in a series held by the Expert Odours Group of the Commission of the European Communities, COST 681 programme.

Its objectives:- to review the present state of knowledge concerning the measurement and control of ammonia and odours emissions from livestock production. To make recommendations for future research needs and to continue the programme of collaboration.

The workshop was successful and achieved its objectives. The papers which form these proceedings set out the present state of knowledge on this very complex subject.

One of the main tasks of the COST 681, Expert Odours Group was the harmonisation of olfactometric measurement. The work on odour concentration has been accomplished following earlier workshops held at Silsoe, UK 1985, Uppsala, Sweden 1987 and Zurich, Switzerland 1988 (1).

Efforts have also been made to harmonise odour intensity measurements. At the workshop in Zurich, a proposal of the Expert Group did not meet sufficient support from relevant experts.

During the final workshop at Silsoe, a small group met to seek agreement on odour intensity measurements and to report to the workshop. The presented papers however, are not available for the proceedings. Moreover, there was again insufficient agreement on the proposed guidelines. As a result, the Expert Odours Group propose to abandon further attempts to harmonise odour intensity measurements.

The organisers of the workshop are grateful to the Director of the Agriculture and Food Research Council's Institute of Engineering Research, Professor John Matthews, for the magnificent facilities and excellent hospitality. Also to V C Nielsen who set out the programme and made all the local arrangements for the workshop, and to Ir J H Voorburg who was the founder of the Expert Odours Group and presided over the arrangements and workshop. Mr G J Monteny provided the Expert Group with summary of the workshop papers at the final session which made a significant contribution to the discussion. Finally, our thanks to Mr P L'Hermite of Waste Recycling DG 12 for his support throughout the series of workshops.

V C Nielsen
J H Voorburg

References

Odour Prevention and Control of Organic Sludge and Livestock Farming:
Nielsen V C, Voorburg J H and L'Hermite P (eds) 1986, Elsevier Applied Science Publishers, London

Volatile Emissions from Livestock Farming and Sewage Operations,
Nielsen V C, Voorburg J H, L'Hermite P (eds) 1988, Elsevier Applied Science Publishers, London.

Measurement of Odour Emissions, Concerted Action, Treatment and Use of Organic Sludge and Liquid Agricultural Wastes, Cost Project 681, SL/130/89, XII/ENV/2/89.

CEC Rue de la Loi 200, B-1049, Brussels.

Hangartner M J, Hartung J, Paduch M, Pain B F and Voorburg J H, 1989. Improved Recommendations on Olfactometric Measurements. Environmental Technology Letters, 110, 231-236. Publications Division Selper Ltd.

CONTENTS

SESSION V –   AMMONIA AND ODOUR LOSSES FROM GRASSLAND

SESSION VI –   DEVELOPMENTS IN ODOUR MEASUREMENTS

Session I

AMMONIA AND ODOUR EMISSION FROM LIVESTOCK PRODUCTION

Chairperson : J.H. VOORBURG

RELATIONSHIPS BETWEEN ODOUR AND AMMONIA EMISSION DURING AND FOLLOWING THE APPLICATION OF SLURRIES TO LAND

B. F. PAIN and T. H. MISSELBROOK

AFRC Institute of Grassland and Environmental Research, Hurley, Maidenhead SL6 5LR, Berkshire, UK

Summary

Micrometeoroglogical methods and a system of small wind tunnels were used in the measurement of odour and ammonia emission following the application of pig slurries to land. Ammonia concentration in air was determined by absorption in acid and odour concentration by dynamic dilution olfactometry. For untreated slurries, close correlations were established between the rate of odour and ammonia emission both during and following application. Similar results were obtained for total odour and ammonia emission following application. No such relationships were established for slurries treated aerobically, anaerobically or acidified prior to application.

1. INTRODUCTION

In common with many other member states, there is in the UK increasing concern about the impact of livestock farming on the wider environment. Spreading slurries on land is a major source of complaints from the public about odours and of emissions of ammonia to the atmosphere. The latter are not only known to be associated with adverse environmental effects but can also result in significant reductions in the value of slurries as fertilizers.

In the field, similar methodology can be used in the measurement of both odour and ammonia emission. Due to the complex chemical composition of odours, olfactometric techniques are used in the determination of odour concentration in air rather than the more rapid, more precise chemical analyses used for ammonia. Although it is known that ammonia is not a major constituent of the perceived odour from slurry[1,2], other workers[3,4] have found good correlations between odour and ammonia emission from slurry storage tanks and from land following application of either pig or cattle slurry. It is feasible that the factors which influence the rate of ammonia volatilization also influence the rate of emission of volatile compounds responsible for odours. The advantages of establishing such relationships are that they could allow odour emission to be estimated from measurements of ammonia emission and for similar strategies to be used to reduce both types of emission.

The aim of the current study was to examine relationships between odour and ammonia emission during or following the application of pig slurries to land. Data were taken from a series of field experiments designed primarily to investigate reduction of emissions by alternative methods of landspreading or by treating slurry prior to application.

2 MATERIALS AND METHODS

Odour collection and measurement

Odorous air was collected in 60 l Teflon bags by means of metal bellows pumps and PTFE tubing. Measurements of odour concentration (the number of dilutions until 50% of a panel of eight people can just perceive an odour) were made within 24 hours of sampling. Measurements were made using two dynamic dilution olfactometers (Project Research, Amsterdam) with a forced choice type of presentation. Each olfactometer had two sniffing ports, one containing clean, odourless air and the other containing the diluted odour sample. Panellists indicated *via* a keyboard which port they considered to contain the odorous air. Each sample was presented to the panel at six dilution steps, each step differing from the next by a factor of x 2, and the range of six dilutions was repeated three times. Equipment and procedures used were as recommended by Hangartner et al.[5]. The odour threshold concentration was calculated by performing a regression of a z-score transformation of panel responses[6] against $\log_{10}$ concentration.

Ammonia sampling

Ammonia concentration in air was measured by drawing the air at 5 l min^{-1} through absorption flasks fitted with a scintered glass distribution head. The ammonia was trapped in orthophosphoric acid (0.002 M) and analysed in the laboratory by a modified Bertholot reaction[7].

Emission during slurry application

Pig slurry was applied to grassland using five different machines. Measurements of odour and ammonia emission during slurry application were made by means of a large sampling frame fitted to the front of a Land Rover[8]. The Land Rover followed each machine in the wake of the slurry plume. The frame had a cross-sectional area of 40 m^2 and supported 16 ammonia absorption flasks and 5 odour sampling tubes. Pumps and odour bags were located in the back of the Land Rover. Total emission during spreading was calculated by multiplying the ammonia or odour concentration by the volume of air passing through the frame. The volume of air passing through the frame was determined as:

air volume (m^3) = (travel speed [m s^{-1}] + wind speed [m s^{-1}]) x cross-sectional area of frame (m^2) x spreading time (s)

Emission following application

A micrometeorological method and a system of small wind tunnels were used in experiments to determine odour and ammonia emission following application of pig slurry.

The micrometeorological method as outlined by Pain and Klarenbeek[8] uses the theory described by Denmead[9] in which the flux of gas from the soil surface of a circular plot is estimated from the vertically integrated product of wind speed and concentration divided by the radius of the plot or the fetch. Ammonia concentrations and wind speeds were measured at 6 heights at the centre of each plot. Odour concentrations were measured at just one pre-determined height, termed ZINST, from which the surface flux could be deduced[10]. These methods were used to compare odour and ammonia emissions following slurry application by three different machines viz. a conventional vacuum tanker, a shallow injector and a deep injector.

A system of small wind tunnels[11] was used in experiments to investigate the effect on odour and ammonia emission of diluting slurry, irrigation with clean water after application, acidification, biological treatment or a commercially available slurry additive. In all experiments slurry was spread on 0.5 x 2 m plots which were then covered by a wind tunnel through which the wind speed was controlled at 1 m s^{-1} and continuously recorded, such that the volume of air passing through each tunnel was known. Samples of the air leaving each tunnel were collected for odour concentration measurement. The concentration of ammonia in the air entering and leaving each tunnel was measured using absorption flasks. Emission of odour and ammonia was calculated as the product of the mean concentration and the volume of air drawn through the tunnel for a given period.

Experiment		% DM	pH	% NH_4^+-N
During		2.9	7.6	0.26
Micromet.		3.4	7.6	0.27
Wind tunnel	1	2.2	7.6	0.23
" "	2	3.3	7.6	0.32
" "	3	2.6	7.5	0.19
" "	4	10.5	7.6	0.35
" "	5	5.1	7.5	0.42
" "	6	4.6	7.6	0.40

Table 1: Some characteristics of the untreated slurries used in the experiments

Slurry
Pig slurries from farms in the UK and The Netherlands were used in these experiments. Dry matter content (% DM), pH and ammonium nitrogen content (% NH_4^+-N) of untreated slurries are given in Table 1.

3. RESULTS AND DISCUSSION
Emission during slurry application
The mean rates of emission of odour and ammonia during slurry application for five different machines are given in Table 2 and values for total emission in Table 3. The differences in emission between the five machines were similar for odour and ammonia and by pooling all the data from 3 replicate experiments with each machine, a significant correlation was obtained (r = 0.851) for rates of emissions. The regression equation was calculated as:
odour emission rate (OU s^{-1}) = 308.2 x ammonia emission rate (mg NH_3-N s^{-1})
The standard error of the slope was 34.8 and percentage variance accounted for was 79.2.

Emissions following application of untreated slurries
Previous studies[4,8] have shown that the rates of emission of odour and ammonia show a similar decline with time after application (Figure 1). Rates of emission are

Machine	Odour emission rate 10^3 OU s^{-1}	Ammonia emission rate rate mg NO$_3$-N s^{-1}
1	7.87 (2.27)	38.4 (11.5)
2	3.30 (2.43)	5.3 (1.9)
3	1.05 (0.78)	5.2 (6.1)
4	2.67 (1.63)	1.2 (1.2)
5	30.52 (27.5)	104.9 (38.0)

( ) = Standard error of mean

Table 2: Mean rate of odour and ammonia emission during slurry application with 5 different types of machine

Machine	Total odour emission 10^3 OU m^{-3} slurry	Total ammonia emission g NH$_3$-N m^{-3} slurry
1	349	1.68
2	133	0.21
3	35	0.16
4	182	0.09
5	6520	22.48

Table 3: Total odour and ammonia emission during slurry application

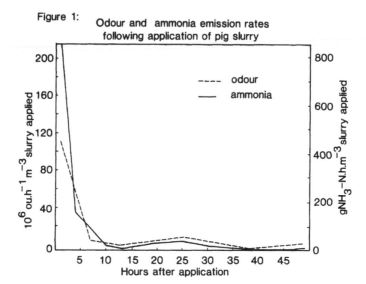

Figure 1: Odour and ammonia emission rates following application of pig slurry

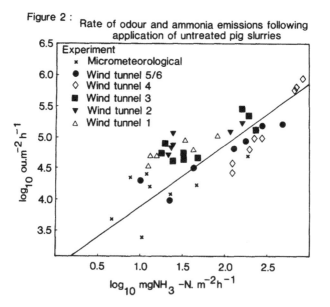

Figure 2 : Rate of odour and ammonia emissions following application of untreated pig slurries

initially high but fall rapidly in the first few hours and then show small diurnal fluctuations. In this study a similar pattern was recorded and, for each of the individual experiments, a good correlation was obtained between the rates of odour and ammonia emission from the untreated slurries. The regression lines (Table 4) differed significantly between experiments. However, a very good correlation was obtained when data from untreated slurries in all of the experiments was combined (Figure 2).

Expt.	Correlation coefficient (r)	Equation	SE of slope	% variance accounted for
Mast	0.851	y = 233 x + 7592	47.9	69.4
WT1	0.888	y = 895 x + 35610	232.0	73.5
WT2	0.855	y = 626 x + 64371	190.0	66.3
WT3	0.816	y = 1037 x	160.0	54.7
WT4	0.989	y = 1082 x - 128679	61.4	97.5
WT5/6	0.951	y = 421 x	41.0	85.5
Combined	0.917	y = 806 x	41.7	84.0

Where y is odour emission rate (OU m^{-2} h) and x is ammonia emission rate (mg NH_3-N m^2 h^{-1})

Table 4: Regression equations for emissions following application of untreated pig slurries

Total emissions of odour and ammonia from untreated slurries at the end of each experiment are given in Table 5. A significant correlation was obtained (r = 0.842) by combining data from all of the experiments.

Experiment		Duration of experiment (h)	Total odour emission 10^3 OU m^{-2}	Total NH_3 emission mg NH_3-N m^{-2}
Micrometeorological:				
	1	48	449	1514
	2		406	242
	3		179	57
Wind tunnel	1	28	1427	554
" "	2	52	4478	1687
" "	3	48	3340	2197
" "	4	48	6157	11350
" "	5	48	1566	715
" "	6	48	1785	957

Table 5: Odour and ammonia emissions from untreated slurry following application

Emissions following application of treated slurries

Results are given in Table 6, of experiments measuring emissions from aerobically treated, anaerobically digested and acidified slurries following application to land. In each case the treatment affected the resulting emissions of odour and ammonia differently. Aerobic treatment resulted in reduced odour emission but increased ammonia emission. Anaerobic digestion resulted in reduced odour emission with no significant effect on ammonia emission. Acidification reduced ammonia emission and possibly increased odour emission. No significant correlations were obtained between odour and ammonia emissions for each of the treated slurries.

| Treatment | Total emission/m^2 over duration of experiment | | | |
| | Untreated slurry | | Treated slurry | |
	10^3 OU	mg NH_3-N	10^3 OU	mg NH3-N
Aeration	4479	1687	1330	2981
Aeration	3341	2197	1673	3330
Digestion	1397	1184	363	927
Digestion	3972	2154	654	2404
Digestion	649	1463	164	1713
Digestion	606	3396	183	1745
Acidification:	pH 7.4		998	1379
	pH 6.1		1325	187
	pH 5.1		1080	47
	pH 3.6		1704	0

Table 6: Emissions of odour and ammonia following application of treated pig slurries

4. CONCLUSIONS

1. Significant correlation was found between odour and ammonia emission during the application of untreated pig slurries.
2. Good positive linear relationship was found between odour and ammonia emission following application of untreated pig slurries, although the slopes of the regression lines differed in different experiments.
3. No such correlation was obtained for treated slurries, thus treatments designed to reduce odour emission would not necessarily reduce ammonia emission, and vice versa.

REFERENCES

(1)   WILLIAMS, A. G. (1984) Indicators of piggery slurry odour offensiveness. Agricultural Wastes **10**, 15-36.

(2)   SPOELSTRA, S. F. (1980) Origin of objectionable odorous components in piggery wastes and the possibility of applying indicator components for studying odour development. Agriculture and Environment **5**, 241-260.

(3)   KOWALEWSKY, W. H., SCHEU, R. & VETTER, H. (1980) Measurement of odour emissions and immissions. In: Effluents from livestock (Gasser, J. K. H. ed.). Applied Science Publishers, London.

(4)   PAIN, B. F., REES, Y. J. & LOCKYER, D. R. (1988) Odour and ammonia emissions following the application of pig or cattle slurry to the land. In: Volatile emissions from livestock farming and sewage operations (Nielsen, V. C., Voorburg, J. H. & L'Hermite, P. eds.). Elsevier Applied Science Publishers, London.

(5)   HANGARTNER, M., HARTUNG, J., PADUCH, M., PAIN, B. F. & VOORBURG, J. H. (1989) Improved recommendations on olfactometric measurements. Environmental Technology Letters **10**, 231-236.

(6)   GUILDFORD, J. P. (1954) Psychometric Methods. McGraw Hill, New York.

(7)   WEATHERBURN, M. W. (1967) Phenol hypochlorite reaction for determination of ammonia. Analytical Chemistry **39**, 971-974.

(8)   PAIN, B. F. & KLARENBEEK, J. V. (1988) Anglo-Dutch experiments on odour and ammonia emissions from landspreading livestock wastes. Research Report 88-2. Instituut voor Mechanisatie, Arbeid en Gebouwen, Wageningen, Netherlands.

(9)   DENMEAD, O. T. (1983) Micrometeorological methods for measuring gaseous losses of nitrogen in the field. In: Gaseous loss of nitrogen from plant-soil systems (Freney, J. R. & Simpson, J. R. eds.). Martinus Nijhoff, The Hague, NL.

(10)   WILSON, J. D., THURTELL, G. W., KIDD, G. E. & BEAUCHAMP, E. G. (1982) Estimation of rate of gaseous mass transfer from a surface source plot to the atmosphere. Atmospheric Environment **16**, 1861-1867.

(11)   LOCKYER, D. R. (1984) A system for the measurement in the field of losses of ammonia through volatilisation. Journal of the Science of Food and Agriculture **35**, 837-848.

ECOLOGICAL EFFECTS OF AMMONIA

J.G.M. ROELOFS and A.L.F.M. HOUDIJK

Dept. of Aquatic Ecology and Biogeology
Catholic University
Toernooiveld
6525 ED Nijmegen, The Netherlands

Summary
Considering the effects of airborne ammonia on ecosystems one can
distinguish between direct and indirect effects. Direct effects on
vegetation normally occur only in cold climates like the northern part
of Scandinavia or in cold winters. Typical symptoms of these ammonia
damages are the red or reddish brown colouring of coniferous trees. It
is caused by a too low ammonia detoxification capacity of plants at
low temperatures. In a mild climate, however, trees normally recover
during summertime. More widespread and serious are the indirect effects
of ammonium deposition on ecosystems. In weakly buffered ecosystems a
high deposition of ammonium leads to acidification and nitrogen enrichment of the soil. As a consequence many plant species characteristic
of poorly buffered environments disappear. Among the acid tolerant
species there will be a competition between slow growing plant species
and fast growing nitrophilous grass or grass-like species. This process
contributes to the often observed change from heath- and peatlands
into grasslands.
    In forest ecosystems a high input of ammonium leads to leaching
of K^+, Mg^{2+} and Ca^{2+} from the soil, often resulting in increased ratios
of NH_4^+ to K^+ and Mg^{2+} and/or Al^{3+} to Ca^{2+} in the soil solution. Field
investigations show a clear correlation between these increased ratios
and the condition of Pinus nigra var. maritima (Ait.) Melville, Pseudotsuga menziesii (Mirb.) Franco and Quercus robur L. Ecophysiological
experiments proved that increased ratios of NH_4^+ to K^+ inhibit the
growth of symbiotic fungi and the uptake of potassium and magnesium by
the root system. At high NH_4^+/K^+ and Al^{2+}/Ca^{2+} ratios there is a net
flux of Mg^{2+} and Ca^{2+} from the root system to the soil solution.
    Other experiments proved that coniferous trees - and most likely
other plant species, too - take up NH_4^+ by the needles, resp. leaves
and compensate for this by excreting K^+ and Mg^{2+}.
    This combination of effects often results in potassium and/or
magnesium deficiencies, severe nitrogen stress, and as a consequence
premature shedding of leaves or needles. Furthermore the trees become
more susceptible to other stress factors such as ozone, drought, frost
and fungal diseases, as well as insect calamities.

1. THE CHANGE OF HEATHLANDS INTO GRASSLANDS

The most obvious phenomenon in many heathlands during the last decades
is the changing from heathland into grassland (Heil, 1984; Heil and Diemont,
1983; Roelofs et al., 1984; Roelofs, 1986). Particularly Molinia caerulea
(L.) Moench and Deschampsia flexuosa (L.) Trin. expand strongly, at the
expense of Calluna vulgaris (L.) Hull and other heathland species.

    In order to estimate whether this phenomenon is related to changes in
the physical-chemical environment, 70 grass-dominated and heather-dominated
heathlands have been investigated (Roelofs, 1986). Many parameters such as
the pH, showed hardly any differences. However, the nitrogen levels in
grass-dominated heathlands appeared to be much higher (Table 1).

Table 1: The pH (H₂O) and average nutrient concentrations in the soil-solution of 70 investigated heathlands.

Species	Coverage	pH (H₂O)	μmoles kg⁻¹			
			NH_4^+	NO_3^-	PO_4^{3-}	K^+
Erica tetralix L.	> 60 %	4.1	55	0.0	4.0	37
Calluna vulgaris (L.) Hull	> 60 %	4.1	84	1.4	4.4	46
Molinia caerulea (L.) Moench	> 60 %	4.2	248	17.2	4.7	88
Deschampsia flexuosa (L.) Trin.	> 60 %	4.1	429	29.0	6.0	182

Both in grass-dominated and heather-dominated heathlands the ammonium levels were 10-20 times higher than the nitrate levels. Investigations clearly show that a major part of the nitrogen originates from atmospheric deposition. Under natural conditions this atmospheric nitrogen deposition is only a few kg.ha⁻¹yr⁻¹. At the present time in the Netherlands the deposition on heathlands often varies between 20 and 60 kg.ha⁻¹yr⁻¹; 60-90% as ammonium-sulphate.

2. SOIL ACIDIFICATION

Although heathland soils are often acidic by nature, there are often certain spots or areas where, due to natural causes (loamy places, a calcareous underground, upwelling deeper groundwater) or to human activities (digging, cattle drinking-places) the soil has become slightly buffered and thus less acidic (Roelofs et al., 1984). Here plant species occur which are restricted to these slightly buffered, less acidic sediments (Table 2).

Table 2: The distribution of some plant species from heathlands in relation to the soil pH.

Species	n	pH (H₂O)		
		mean	min.	max.
Erica tetralix L.	> 10	4.1		
Calluna vulgaris (L.) Hull	> 10	4.1	4.0	4.3
Molinia caerulea (L.) Moench	> 10	4.2	3.8	4.7
Polygala serpyllifolia Hose	> 10	4.5	4.1	5.7
Lycopodium inundatum L.	> 10	4.6	4.4	4.9
Pedicularis sylvatica L.	> 10	4.7	4.2	5.9
Thymus serpyllum L.	> 10	5.1	4.7	5.6

These plant species like Thymus serpyllum L. and Pedicularis sylvatica L. never occur on sediments with a pH value as low as 4.1. The deposited ammonium at these slightly buffered locations is transformed into nitrate very quickly by nitrification, which causes acidification of the soil (Van Breemen et al., 1983; Roelofs et al., 1984).

Laboratory experiments with artificially buffered heathland soils show that nitrification stops or is strongly inhibited in this type of soil at pH 4.1 (Roelofs et al., 1985). This appeared also to be case for the average pH-value in both grass-dominated and heather-dominated heathlands, which indicates that the pH in heathlands is probably determined by the nitrification limit. The final result of high NH_4^+ deposition levels is that the differences in pH disappear and thus also the plant species of slightly buffered locations. A poor plant community remains, consisting of only a few acid resistant species.

3. NITROGEN ENRICHMENT

If the soil on which ammonium is deposited acidifies, a strong accumulation of nitrogen occurs in the upper soil layer, because ammonium is bound much more strongly to the soil absorption complex than nitrate. When there is competition between heather species such as Erica tetralix L. and Calluna vulgaris (L.) Hull and grasses such as Molinia caerulea (L.) Moench the grasses profit from these higher nitrogen levels (Scheikh, 1969; Heil and Diemont, 1983; Berendse and Aerts, 1984; Heil, 1984; Roelofs et al., 1984; Roelofs, 1986). Field fertilisation experiments have shown that nitrogen enrichment indeed stimulates the development of grasses in heathlands (Heil and Diemont, 1983).

However, the problem with these field fertilisation experiments is that the high atmospheric nitrogen deposition was not taken into account. For this reason, experiments were carried out in a greenhouse. A number of small heathlands were created, using undisturbed, natural heathland soils. Precipitation experiments during one year showed that the biomass development of the grasses Agrostis canina L. and Molinia caerulea is not influenced by the acidity of the precipitation (Figure 1). If the precipitation contained ammonium sulphate, a strong increase in biomass with increasing NH_4^+ deposition was observed. The chosen annual ammonium deposition was comparable with the real field deposition. The increase in biomass of Molinia was the strongest between 1.4 and 2.8 $kmol.ha^{-1}yr^{-1}$ (= 20 and $\overline{40\ kg.ha^{-1}yr^{-1}}$).

The results of these experiments show that the NH_4^+ deposition level in the Netherlands (20-60 $kg.ha^{-1}yr^{-1}$) cause a marked increase in biomass of the two investigated grass species. For this reason it can be concluded that the high atmospheric nitrogen enrichment is a main cause for changes from heather-dominated into grass-dominated heathlands.

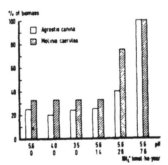

Fig. 1: The relation biomass development of *Agrostis canina* and *Molinia caerulea* on natural heathland soil during a one year treatment with precipitation with different pH and ammonium concentrations in a greenhouse.

4. AMMONIUM DEPOSITION AND THE CONDITION OF FORESTS

The condition of the Dutch forests is alarming. A recent investigation by the Dutch State Forest Service reveals that 49.9% of the forest stands show a decreased vitality (Anonymus, 1989). The geographical pattern of the damage does not fit in very well with the occurrence of well-known pollutants as SO_2, NO_x and O_3 (Den Boer, 1986). The situation is most critical in the south-eastern part of the country. Here nitrogen deposition in forest stands is very high and about 10-20 times the natural supply of 5-10 $kg.ha^{-1}$

yr^{-1}. Due to the filtering action of the tree canopies deposition of gaseous ammonia, sulphur, dioxyde and ammonium sulphate is considerably higher in the forests than in the surrounding meadows (Van Breemen et al., 1983; Nihlgård, 1985; Roelofs et al., 1985; see also Table 3).

Table 3: The average chemical composition of precipitation in open plots and throughfall in *P. nigra* forests in three different regions of the Netherlands during 1984 (μM/l).

	H^+	NH_4^+	K^+	Na^+	Ca^{2+}	Mg^{2+}	NO_3^-	Cl^-	SO_4^{2-}
North-West (Terschelling)									
open	100	65	25	308	47	62	48	370	72
throughfall	400	59	345	6700	460	1120	110	8000	860
South (Heeze)									
open	45	130	19	60	43	17	54	73	70
throughfall	6	1060	170	310	200	113	216	350	760
South-east (Venray)									
open	2	200	20	40	30	15	49	50	90
throughfall	1	2421	216	175	278	100	147	462	1400

Several authors mention a relation between agricultural activities and the condition of pine trees (Hunger, 1978; Janssen, 1982; Roelofs et al., 1985). Mainly four types of damage can be observed:

a) red or brown colouring of the needles of all year classes;
b) yellowing of the needles; the older needles more frequently;
c) yellowing of the youngest needles which is most pronounced at the base of the needles;
d) the occurrence of fungal or insect diseases.

All these damages can be related to high or disturbed nitrogen budgets (Roelofs et al., 1985; Roelofs, 1986; van Dijk and Roelofs, 1987). The first type of damage mainly occurs in the neighbourhood of ammonia sources like farms or fields dressed with animal slurry (Janssen, 1982). It is caused by a combination of low temperature (frost) and high ammonia concentration in the air, probably as a result of a too low ammonia detoxifying capacity of the trees at low temperature (Van der Eerden, 1982).

The second type of damage, the yellowing of the needles, is related to potassium and/or magnesium deficiencies and are very significantly correlated with disturbed nitrogen budgets in both air and forest soil (Roelofs et al., 1985; Roelofs, 1986).

The third type of damage, the yellowing of the youngest needles, is strongly correlated with extremely high arginine levels in the needles, high ammonium concentrations in the precipitation and disturbed nitrogen budgets in the soil solution (Van Dijk and Roelofs, 1987).

The last mentioned type of damage, the fungal and insect diseases, may be related to the disturbed nutrient balance in the plant tissue. Investigations in P. nigra forests have shown that all trees infected with the fungus Sphaeropsis sapinea (Fr.) Dyko and Sutton had significantly higher nitrogen levels in the needles compared to non-infected healthy trees (Roelofs et al., 1985).

5. AMMONIA AND THE DISTURBED NUTRIENT BALANCE IN THE TREES

In fact there are two ways in which ammonia can contribute to the observed nutrient disbalance in trees. At first by ammonia/ammonium uptake by the leaves and secondly because a high deposition of NH_3/NH_4^+ can cause a disturbed balance in the soil.

6. THE ROLE OF AMMONIUM IN RELATION TO THE CANOPY-ION EXCHANGE

On locations with relatively low ammonium deposition levels, the ammonium concentration in the throughfall is even lower compared to the open-air bulk precipitation (Table 3). Persson and Broberg (1985) found the same in pine forests in the lake Gårdsjön area (Sweden). On the locations with very high ammonium deposition levels sulphate is completely compensated by ammonium in the bulk precipitation, but only 75-85% in the throughfall. All these observations may indicate that the needles take up ammonium, which also has been observed by Lovett and Lindberg (1984) for mixed oak forests. This implicates that, when total nitrogen-deposition in forests is calculated from throughfall analysis, there might be a strong underestimation. Cation exchange experiments with needles of Corsican pine (Pinus nigra var. maritima (Ait.) Melville) and Douglas fir (Pseudotsuga menziesii (Mirb.) Franco) in artificial ammonium sulphate containing rain proved that the needles take up large amounts of ammonia and compensate by excreting equivalent quantities of potassium, magnesium and calcium (Table 4).

Table 4: Cation exchange of needles in acidic (pH = 4.8 ± 0.4) artificial rainwater containing 100 μM sodium chloride (blanc) or 100 μM sodium chloride + 250 μM ammonium sulphate ($\mu mol \cdot g^{-1}$ DW24h^{-1}).

Cation	NH_4^+		K^+		Mg^{2+}		Ca^{2+}	
	blanc	$+NH_4^+$	blanc	$+NH_4^+$	blanc	$+NH_4^+$	blanc	$+NH_4^+$
Pinus nigra	0.0	3.5	−1.0	−2.0	−0.1	−1.1	0.0	−0.8
Pseudotsuga menziesii	0.0	2.7	−0.3	−0.8	−0.1	−0.3	−0.2	−0.6

This process proceeds continuously in the course of time (Roelofs et al., 1985). Already at moderate ammonium concentration the leaching of Mg^{2+} by the needles can be more than ten times higher compared to acid artificial rain without ammonium.

7. AMMONIUM DEPOSITION AND A DISTURBED NUTRIENT BALANCE IN THE SOIL

Most of the Dutch forest stands are planted on acidic, nutrient poor heathland soils. It is well-known that nitrification, and thus acidification is possible in acidic forest soils (Van Breemen et al., 1982; Klein et al., 1983; Kriebitzsch, 1978). Whether deposition of ammonium sulphate on acidic forest soils will result in strong acidification depends on the type of the forest soil. Kriebitzsch(1978) who conducted nitrification experiments in many types of acidic forest soils, divided them into groups: A, B, C and D. In the soils belonging to group A there was no nitrification. In the soils belonging to groups B and C there was partial nitrification and in those belonging to group D there was total nitrification. The investigations in this study showed that heathlands and Pinus soils mainly belong to group A. Field studies in the Netherlands in Pinus nigra and Pseudotsuga menziesii forests on former heathland soils showed indeed only partial or no nitrification. The nitrate levels were low, whereas the ammonium levels were high (Roelofs et al., 1985; Roelofs, 1986; Table 5)

The soils of healthy, moderately damaged and severely damaged forests had on an average a pH (H_2O) of 4.1, which indicates that also in this type of forest soil the pH is determined by the nitrification limit. In this type of soil a high ammonium deposition level leads to accumulation of NH_4 and leaching of K, Mg and Ca from the soil. As a result the NH_4/K, NH_4/Mg and the Al/Ca ratios increase. It is well-known that increased NH_4 to K and Mg ratios inhibit K and Mg uptake (Jacobs, 1958; Mulder, 1956). In both Pseudotsuga menziesii and Pinus nigra forests the NH_4/K and NH_4/Mg ratios are relatively low in healthy forests and significantly higher in severely

Table 5: pH and chemical composition of soil-distilled water extracts (1:3) of A) healthy, B) moderately damaged and C) severely damaged *Pinus nigra* and *Pseudotsuga menziesii* forests (μmoles kg^{-1} dry soil).

		pH (H$_2$O)			NH$_4^+$	NH$_4^+$ (KCl)*	NO$_3^-$	K$^+$	Mg^{2+}	Ca^{2+}	Al^{3+}
	n	mean	min.	max.	mean	mean	mean	mean	mean	mean	mean
Pinus nigra											
A)	20	4.1	3.5	4.6	334	687	271	137	77	153	191
B)	16	4.0	3.4	4.9	384	751	130	47	45	128	158
C)	20	4.1	3.7	4.4	509	1346	227	60	26	43	183
Pseudotsuga menziesii											
A)	10	4.1	3.9	4.4	245	499	164	89	60	106	214
B)	10	4.1	3.8	4.3	562	733	153	67	48	69	211
C)	11	4.3	4.0	4.6	692	1240	157	67	22	36	211

* 0.5 M KCl extract.

Table 6: The ratios of some nutrients in soil extracts of A) healthy, B) moderately damaged and C) severely damaged *Pinus nigra* and *Pseudotsuga menziesii* forests (mol/mol).

		NH$_4^+$/K$^+$			NH$_4^+$/Mg^{2+}			Al^{3+}/Ca^{2+}		
	n	mean	min.	max.	mean	min.	max.	mean	min.	max.
Pinus nigra										
A)	21	4.7	0.5	14.0	6.4	1.1	24.3	2.0	0.4	5.6
B)	17	9.2	0.8	36.8	10.0	1.8	26.3	1.3	0.2	2.8
C)	21	11.3	1.9	51.8	22.1	1.6	57.2	5.5	1.7	16.7
Pseudotsuga menziesii										
A)	10	3.8	0.5	11.8	4.5	0.6	10.0	6.6	0.8	40.9
B)	10	8.7	1.5	31.2	19.3	2.0	51.1	8.9	0.7	46.7
C)	11	18.2	3.8	64.5	47.6	7.9	118.0	15.6	1.4	54.0

damaged forests and particularly in Pseudotsuga forests (Table 6), far above the critical value for root damage (Ulrich, 1983).

Apart from premature shedding of needles as a result of nutrient deficiencies and severe nitrogen stress, the forests become more susceptible to other stress factors such as O$_3$, drought, frost and fungal diseases.

ACKNOWLEDGEMENT

We wish to thank Mrs. M.G.B. van Kuppeveld-Kuiper for typing this manuscript.

REFERENCES

ANONYMUS (1989). De vitaliteit van het Nederlandse bos. Ministry of Agriculture and Fisheries.
BERENDSE, F. and AERTS, R. (1984). Competition between Erica tetralix L. and Molinia caerulea (L.) Moench as affected by the availability of nutrients. Oecologia Pl.5, 1-13.
BREEMEN, N. VAN, BURROUGH, P.A., VELTHORST, E.J., DOBBEN, H.F., WIT, T. DE, RIDDER, T.B. and REYNDERS, H.F.R. (1983). Soil acidification from atmospheric ammonium sulphate in forest canopy throughfall. Nature 229, 548-550.
DIJK, H.F.G. VAN and ROELOFS, J.G.M. (1987). Effects of airborne ammonium on the nutritional status and condition of Pine needles. Proc. COST Workshop on direct effects of dry and wet deposition on forest ecosystems. In particular canopy interactions. Lökeberg, Sweden, 19-23 Oct. 1986.
EERDEN, L.J.M. VAN. (1982). Toxicity of ammonia to plants. Agriculture and Environment 7, 223-235.
HEIL, G.W. (1984). Nutrients and the species composition of heathlands. Ph.D. Thesis, Univ. of Utrecht 1984.

HEIL, G.W. and DIEMONT, W.M. (1983). Raised nutrient levels change heathland into grassland. Vegetatio 53, 113-120.

HUNGER, W. (1978). Über Absterbeerscheinungen an älteren Fichtenbeständen in der Nähe einer Schweinemastanlage. Beitr. Forstwirtsch. 4, 188-189.

JACOB, A. (1958). Magnesia, der fünfte Pflanzenhauptnährstoff. Stuttgart, Enke.

JANSSEN, TH.W. (1982). Intensieve veehouderij in relatie tot ruimte en milieu. Dutch State Forest Service, Utrecht.

KLEIN, T.M., KREITINGER, J.P. and ALEXANDER, M. (1983). Nitrate formation in acid forest soils from the Adirondacks. Soil. Sci. Soc. Am. J. 47, 506-508.

KRIEBITZSCH, W.U. (1978). Stickstoffnachlieferung in sauren Waldböden Nordwest-deutschlands. Scripta Geobotanica. Göttingen, Goltze.

LOVETT, G.M. and LINDBERG, S.E. (1956). Dry deposition and canopy exchange in a mixed oak forest as determined by analysis of throughfall. J. Appl. Ecol. 21, 1013-1027.

NIHLGÅRD, B. (1985). The ammonia hypothesis: An additional explanation to the forest dieback in Europe. Ambio 14, 2-8.

PERSSON, G. and BROBERG, O. (1985). Nutrient concentrations in the acidified lake Gårdsjön: The role of transport and retention of phosphorus. Ecological Bulletins 37, 158-175.

ROELOFS, J.G.M. (1986). The effect of air-borne sulphur and nitrogen deposition on aquatic and terrestrial heathland vegetation. Experientia 42, 372-377.

ROELOFS, J.F.M., CLASQUIN, L.G.M., DRIESSEN, I.M.C. and KEMPERS, A.J. (1984). De gevolgen van zwavel en stikstofhoudende neerslag op de vegetatie in heide- en heideveenmilieus. In: Zure regen, oorzaken, effecten en beleid. Eds.: E.H. Adema and J. van Ham, p. 134-240. Proc. Symp. Zure regen, 's-Hertogenbosch, 17-18 November 1983. Wageningen, Pudoc.

ROELOFS, J.G.M., KEMPERS, A.J., HOUDIJK, A.L.F.M. and JANSEN, J. (1985). The effect of airborne ammonium sulphate on Pinus nigra var. maritima in the Netherlands. Pl. Soil 84, 45-56.

SCHEIKH, K.H. (1969). The effects of competition and nutrition on the interrelations of some wet-heath plants. J. Ecol. 57, 87-99.

SCHNEIDER, T. and BRESSER, A.H.M. (eds.) (1988). Dutch priority programme on acidification. Rapport nr. 00-06 R.I.V.M. Bilthoven, the Netherlands, 190 pp.

ULRICH, B. (1983). Soil acidity and its relation to acid deposition. In: Effects of accumulation of air pollutants in forest ecosystems, p. 127-146. Eds.: B. Ulrich and J. Pankrath. Dordrecht: D. Reidel Publ. Comp.

THE ROLE OF AMMONIA AS AN ATMOSPHERIC POLLUTANT

H.M. APSIMON, M. KRUSE-PLASS

Air Pollution Group, Imperial College, London SW7 2AZ (UK)

Ammonia is the most prevalent alkaline gas in the atmosphere, and as such plays an important role in atmospheric chemistry, and acid deposition. In the form of ammonium aerosols ammonia can be transported over long distances, thus constituting a pollutant on an international scale. Until a few years ago it was not widely appreciated that acid deposition of for example $(NH_4)_2SO_4$ is potentially more acidifying for soils than strong acid H_2SO_4. Moreover emissions of nitrogen as ammonia (~5 million tons per annum) exceed those of nitrogen as NOx in Europe, with important implications with respect to critical loads.

Agricultural activities are the dominant sources of ammonia, and emissions in Europe have risen along with more intensive farming and use of fertilizers. However the direct emissions from fertilizer applications on arable land are generally small compared with those from livestock farming using the feedstocks produced. Independent assessments of emissions have been undertaken by ApSimon and Kruse and Bell for the UK (1987), and Asman and Buisjman (Asman and Jennsen 1987) for Europe as a whole. These inventories are in fairly close agreement, but the authors of both conclude from modelling studies that they represent an underestimate. Possible reasons for this are considered by Kruse–Plass and ApSimon (Kruse, ApSimon and Bell 1989). We also estimate that ammonia emissions in Europe have increased by about 50% since 1950, with a doubling in some countries such as the Netherlands.

The fate of ammonia emissions in the atmosphere is a complicated one–see figure 1. Released at ground level, some of the ammonia is redeposited before it can diffuse upwards into the atmosphere. This will depend on such factors as the vegetation and soils, and dew and humidity, as well as air concentrations. There are large diurnal variations, and the same area may alternate between acting as a net sink and a net source. Due to the processes indicated in figure 1 both ammonia and ammonium exhibit marked variations in vertical concentration profiles, with NH_3 levels decreasing with height. Typical values of ammonia a few meters above the ground vary from one or two $\mu g.m^{-3}$ to a few tens of $\mu g.m^{-3}$. See for example Harrison et al (Allen, Harrison and Wake, 1988) who undertook measurements at a range of urban and rural sites in eastern England.

Ammonia is a reactive gas, and as it mixes upwards it readily combines with acidic species such as H Cl, HNO_3 and H_2SO_4 forming ammonium aerosols. In this context the ammonia plays a fairly passive role, merely being mopped up onto the products of photo-oxidant processes in the atmosphere. However some ammonia will penetrate up into clouds where it can play a very influential role in cloud chemistry. In particular it affects the oxidation of SO_2 by ozone in cloud droplets, which is important in facilitating the removal of SOx in precipitation. The ozone reaction complements another process of in–cloud oxidation of SO_2, by H_2O_2; but is more effective at higher pH values, reducing sharply as pH falls to values of 4.0 and below. As a soluble gas ammonia has a marked effect on pH, and as a result controls the ozone oxidation rate.

To study the role of ammonia in the transport of ammonia and its interaction with acidic species we have developed a numerical model called MARTA (Modelling the Atmospheric Release and Transport of Ammonia). This model follows a column of air along

18

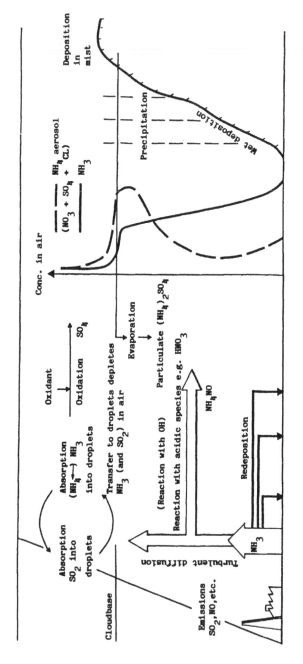

The fate of NH_3 emissions into the atmosphere

calculated trajectories across Europe to specified receptor points. The model considers up to 20 horizontal slices in each air column to reflect vertical profiles of diffusion, moisture and interacting pollutants. Meteorological data along the trajectories on such factors as cloud cover and base of cloud, are taken from synoptic observations.

Our modelling studies on the long–range transport of acidic species have demonstrated the strong feed–back resulting from the presence of ammonia between the different sulphur oxidation processes; with ozone oxidation in clouds increasing(reducing) with lower(higher) availability of H_2O_2. The latter is produced by photo–oxidant activity, increasing with solar radiation and high ozone concentrations.

Four studies undertaken provide illustration of the effects of ammonia on acidic deposition:

i)   The relatively high ammonia emissions from Wales and the cattle rearing regions of Herefordshire provide an explanation for rapid oxidation and deposition of sulphur observed over Wales in an experimental campaign observing SOx fluxes emanating from the industrial Midlands in easterly winds.
(Bamber et al 1983)

ii)   The dense ammonia emissions over the Netherlands can have a significant effect on the oxidation of the polluted air masses resulting from high SO_2 emissions in central and eastern Europe. A reduction in ammonia emissions by the Netherlands could increase the proportion of unoxidised SO_2 vented to the free troposphere in precipitation systems, thus increasing the range over which it is transported.

iii)   Scenarios corresponding to the highest episodes of deposition of both SOx and NH_4 involve high oxidation of the SO_2, in which ammonia emissions often appear to be a controlling factor. Modelling studies imply that, due to the NH_3, overall oxidation is maintained at broadly similar levels despite order of magnitude variations in supply of H_2O_2.

iv)   The presence or absence of ammonia is also highly relevant to the conditions in which very acidic clouds can lead to acid mists over forested hill–tops causing damage to foliage.
(ApSimon and Whitcombe–in press)

Clouds form and evaporate many times for each time they yield precipitation. High concentrations of $(NH_4)_2SO_4$ aerosol are observed emanating from evaporating cloud and also in conditions of high humidity, with implications for visibility. On the basis of half the 50% increase in European ammonia emissions since 1950 giving $(NH_4)_2SO_4$ aerosols with a mean lifetime of 1 to 2 days over Europe the resulting average increase in aerosol burdens would be about 1 $\mu g.m^{-3}$ of sulphate in the lowest 1000 metres of the atmosphere (the approximate depth of the well–mixed boundary layer of the atmosphere). This indicates a significant change, especially in the context of sulphate episodes and degraded visibility.

CONCLUSIONS

It is thus concluded that ammonia is a very important trace constituent in the atmosphere, deserving far more attention with respect to atmospheric chemistry, acidification, aerosol loadings and visibility, and critical loads of nitrogen deposition.

REFERENCES
(1)   ALLEN, A.G., HARRISON, R.M. and WAKE, M.T. 1988. A meso–scale study of the behaviour of ammonia and ammonium. Atmos. Env.22 1347–1353

(2)   APSIMON, H.M., KRUSE, M. and BELL, J.N.B. 1987. Ammonia emissions and their role in acid deposition. Atmos. Env. 21 1939–46

20

(3)   APSIMON, H.M., KRUSE–PLASS, M. and WHITCOMBE, G.. Studies of deposition in hill–cloud. To be published in Il Nuovo Cimento.

(4)   ASMAN, A.H. and JANNSEN, A.J. 1987. Long–range transport modelling of ammonia and ammonium in Europe. EURASAP symposium on Ammonia and Acidification, Bilthoven, The Netherlands.

(5)   BAMBER, D.J. et al. 1983. Air sample flights round the British Isles at low altitudes. Atmos. Env. 18 1777–1790

(6)   KRUSE, M., APSIMON, H.M. and BELL, J.N.B. 1989. Validity and uncertainty in the calculation of an emission inventory for ammonia arising from agriculture in Great Britain. Env. Pollution 56 237–257

Session II

AMMONIA AND ODOUR EMISSIONS FROM BUILDINGS AND STORES

Chairperson : M. PADUCH

INFLUENCE OF HOUSING AND LIVESTOCK ON AMMONIA RELEASE
FROM BUILDINGS

J. Hartung
Institute for Animal Hygiene and Animal Protection,
School of Veterinary Medicine, Bünteweg 17p, 3000 Hanover 71, FRG

Summary

The role of ammonia in soil acidification and in other ecological effects has gained more attention in the last few
years. It is estimated that about 80 % of the ammonia emissions derive from animal husbandry while only 20 % are from
other sources. Calculations show that more than one third of
these ammonia emissions are from buildings and stores. Measures to reduce or prevent ammonia release from buildings
are lowering the temperature and/or the pH-value of the
manure, quick drying of chicken manure, quick removal of
manure and urine in particular to covered storage outside
the building, use of additives to influence chemical and
bacterial processes involved in ammonia formation, dilution
of the manure by adding water, restricted nitrogen feeding.
More information and experience on the different methods for
reducing ammonia emissions is necessary. Ammonia emissions
are losses of valuable nitrogen fertilizer which should be
minimized not only for environmental but also for economical
reasons.

1. Introduction

The contribution of ammonia to total nitrogen deposition and its
role in the acidification of soils and surface waters is now
being recognized, especially in countries with intensive animal
husbandry (1). Table 1 shows a nitrogen balance of the German
agriculture where ammonia emissions represent about 37 % of the
total nitrogen losses (2). It is estimated that more than 90 % of
all ammonia emissions are originating from agriculture and about
80 % from animal production with buildings and stores, landspreading and grazing being the predominant sources.

The amounts of ammonia and other gases which are really emitted
from animal husbandry are not well known, yet (3). In general,
cattle is supposed to produce the greatest amount of the ammonia
followed by pigs and poultry. The other animals are of less importance. In Germany, horses and sheep together come to about the
same amount as poultry. Table 2 gives the amounts of ammonia
emission in the F.R. of Germany calculated for the year 1988 (4).
The calculation is based on the assumption that about 30 % of the
manure nitrogen is lost in form of ammonia. The total emission of
287 kt/y is almost equal to figures given by Buijsman et al. (5)
who estimated 329 kt/y for the F.R. of Germany, whereas Isermann
(2) came to about 500 kt/y. Similar discrepancies are observed in
The Netherlands where the estimations range from 128 kt/y (5) to
220 kt/y (6).

Tab. 1: Nitrogen balance in German agriculture in kg N/ha
agricultural acreage* in 1986 (from (2))

INPUT
mineral fertilizer: 126
imported feed: 47
atmosphere: 30
biological N-fixation: 12
sewage sludge: 3

total   218 (= 2.6 mio. t/y)

OUTPUT
animal products: 28
plant products: 23

SURPLUS   167 (= 77 % of the INPUT, 2 mio. t/y)
N-immobilization/
accumulation in the soil: 47 (= 564 000 t/y -> soil

N-losses: 120 (= 1445 000 t/y
ammonia-emissions: 44 (= 528 000 t/y -> atmosphere
denitrification: 25 (= 300 000 t/y ->   "
leaching: 45 (= 537 000 t/y -> hydrosphere
run off: 5 (=   65 000 t/y ->   "
other losses: 1 (=   15 000 t/y ->   "

NITROGEN EFFICIENCY
total agriculture: 51/218   x   100 % =   23 %
- animal production: 28/167   x   100 % =   17 %
- plant production: 138/190   x   100 % =   73 %
*agricultural acreage 12 mio. ha

Table 2: Ammonia emissions from livestock production in the
F.R. of Germany (4)

livestock	A.U.* in 1000	ammonia-emission kg/A.U./y	t/y	%
cattle	10 112	22	222 461	77
pig	2 612	19	49 627	17
poultry	305	25	7 627	3
horse	368	10	3 676	1
sheep	138	32	4 425	2
total	13 534	-	287 816	100

*A.U. = animal unit = 500 kg living weight

The contributions of the different sections of animal production
to the ammonia emission are given in Table 3. Based on experiences in The Netherlands it is estimated that field application
contributes to more than 50 % to the emission of ammonia from
livestock production followed by buildings and stores (37 %) and
grazing (12 %). A differentiation between buildings and stores is
difficult because in many keeping systems the manure is stored
inside the building, e.g. under the slatted floor, for a longer
period of time.

The results show that there still is a lack of knowledge about
the true amounts which are emitted. On the other hand it is

Tab. 3: Ammonia emissions in kt/y from different areas of animal
keeping (6)

animals	stalls and storage		spreading		grazing		total	
	kt	%	kt	%	kt	%	kt	%
cattle	43	53	71	64	27	19	141	64
pigs	20	24	31	28	-	-	51	23
poultry	19	23	9	8	-	-	28	13
total kt	82		111		27		220	
percent %	37		51		12		100	

agreed that a reduction must take place; this is true not only
for environmental reasons but also from an economical point of
view. Therefore, both the inventory of the emissions should be
improved and measures to reduce the emissions have to be established.

Possibilities and measures to reduce ammonia emissions from buildings are discussed in the following.

2. Influence of temperature, pH-value and drying

Ammonia is formed by bacterial degradation of urea or uric acid
from urine and feces which are just excreted or stored inside of
the animal house. Bacterial activity depends on temperature. The
lower the temperature the lower the activity of the bacteria. The
influence of temperature (7) is demonstrated by seasonal fluctuations of the ammonia-N loss in relation to the aerial temperature in a pig barn (Figure 1). However, a rather high technical
effort is required in order to keep the temperature low during
the summer time under practical farm conditions.

Low pH-values prevent the release of ammonia from water because
most of the ammonia is present in form of ammonium. Miner (8)
showed that increasing pH-values increase the loss of ammonia
from liquid manure. Figure 2 demonstrates the influence of pH-
value and temperature on ammonia emissions. Up to now there is no
procedure available to adjust the pH-value in liquid manure under
practical conditions; experiments are under investigation in
Belgium and in The Netherlands.

Drying of poultry manure by ventilating the manure removal belt
can reduce the ammonia-N loss by a factor of 6 (Table 4). Attention has to be payed to the storing of the dryed manure outside
of the animal house; high ammonia losses can occur by composting
processes.

3. Mechanical removal of the manure

Removing the droppings out of the building is considered a simple
way to reduce ammonia emissions. However, it is important to
remove feces and urine as soon as possible because fresh urine is
a considerable source of ammonia. The good efficiency of reducing

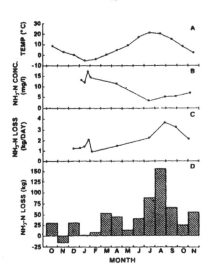

Fig. 1: Seasonal fluctuation in a pig barn. (A) Aerial
temperature. (B) Atmospheric ammonia-N concentration in
the barn. (C) Ammonia-N loss estimated by atmospheric
sampling. (D) Ammonia-N loss estimated from the N mass
budget. (7).

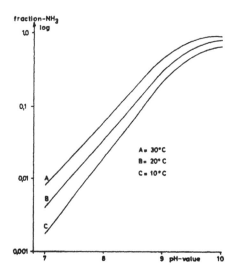

Fig. 2: Fraction of total NHx concentration present as ammonia as
a function of temperature and pH-value (calc. from (8))

Tab. 4: Gaseous nitrogen losses from laying hen cage batteries
with and without ventilating the manure removal belt (9)

manure sample	N-loss		NH_3-N-loss		% NH_3-N
	g	%	g	%	of N-loss
ventilation	4.9	11	0.1	0.2	2.0
no ventilation	8.4	32	1.0	4.0	12.4

aerial ammonia concentrations by removing the manure is demonstrated for poultry manure by Kroodsma et al. (10). Ammonia
concentrations were measured in three different housing systems
for layers. The average ammonia concentration was lowered by
removing the manure by means of a manure belt under the cages
twice a week in comparison to a two-tier step batterie without a
manure belt. The waste removal could be reduced to once a week by
using a drying equipment. However, the aerial ammonia concentration increases considerably during the process of mucking out.
High ammonia emissions occur when the floor is soiled with manure. Therefore the use of surface scrapers seems not to be very
helpful in reducing ammonia release.

4. Addition of water

Kellems et al. (11) found that the volatilization of ammonia from
stored bovine urine was much higher than from the feces itself or
a mixture of both and the addition of water helped to reduce the
ammonia release (Table 5). This may encourage the further development of waste removal systems using flushing.

Tab. 5: Volatilization of ammonia from stored bovine
feces and urine (11)

composition of the manure			ammonia release
feces	urine	water	μgNH_3/h
100 %	-	-	3.15
-	100 %	-	426.0
50 %	50 %	-	120.0
75 %	25 %	-	16.0
75 %	-	25 %	3.4
50 %	-	50 %	6.6
25 %	-	75 %	9.7
5 %	-	95 %	2.2

5. Use of bedding

Other measures to reduce ammonia emission from buildings like the
use of bedding materials (e.g. 12) depend very much on the amount
and maintenance of the bedding. Table 6 shows that bedding can
reduce the ammonia release by a factor of two for poultry and by
a factor of 4 for pigs, respectively. However, these results are
of limited value. It is assumed that solid manure will loose most
of the ammonia during the process of mucking out and during the

storage outside of the building. Thus, the total ammonia volatilization from solid manure does not seem to be considerably lower than that from liquid manure.

Tab. 6: Ammonia emissions from different livestock and from different keeping systems (4)

livestock	keeping/manure system	ammonia emission g/h	kg/year
poultry	liquid manure	4.4	38.8
poultry	bedding/floor	1.9	16.7
pig	liquid manure	0.9	7.5
pig	bedding	0.2	1.7

6. Feed and additives

Promising attempts to lower the nitrogen losses from the manure are N-restrictive feeding and the use of additives which bind the ammonia in the manure. An example is given in Table 7 which compares the excretion of crude protein to different protein supply situations. A reduction of the crude protein intake of about 20 % resulted in a lower nitrogen excretion of about 30 %; the gain was not influenced (13, 14). However, the reduction of the protein intake is limited because of economic reasons, and the success in reducing ammonia release can only be achieved in combination with methods which prevent decomposition of the nitrogen after excretion.

Tab. 7: Comparison of the excretion of crude protein under different protein supplementations in fattening pigs (from (13))

	protein minimum	DLG* recommendation
feed kg	239	241
crude protein:		
intake kg	28.7	35.4
gain kg	13	13
excretion N kg	2.5	3.6

*DLG=German Agricultural Society

Table 8 shows that feeding-rations can influence the pH-value of the manure and consequently the ammonia evolution rates (11). However, only a trend supporting the opinion that low pH-values result in low ammonia evolution rates can be observed because of high deviations of the results presented.

The success of additives in feed and manure is still inconclu-

sive. Numerous positive reports are in opposition to practical
experiences. One of these positive reports from investigations on
feed additives in vitro is given in Table 9 (15). The authors
found similar effects under practical conditions. Suppressing
production of ammonia through chemical litter treatments has had
limited success so far (e.g. 16, 17, 18).

Tab. 8: pH and NH_3 evolution rates from feces and urine mixtures
(50 g feces/50 g urine) collected from cattle receiving
75 % milo, corn and barley rations. Means of n = 28 +
standard error of mean. pH-values differ significantly
($p \leq 0.05$) (11)

ration	average pH	average ammonia (µg/h)
milo	6.78 + .08	2602.03 + 302.56
corn	7.21 + .06	2731.96 + 304.72
barley	7.65 + .08	3182.48 + 241.63

Tab. 9: Ammonia concentration[a] in poultry excreta decomposing
in vitro (15)

Micro Aid added (ppm)	incubation time (h)			overall[b]
	7	14	21	
0	4.74	8.24	15.13	8.71
62	2.82	6.25	12.05	6.82
124	2.70	4.78	8.18	5.03
overall[c,d]	3.43	6.46	11.60	

a=Values are mg ammonia/dl mixture fluid (after adjusting
 initial ammonia concentration)
b=Micro Aid main effect (p<.01)
c=Incubation time main effect (p<.01)
d=Micro Aid x incubation time interaction (p<.04)

7. Discussion

This short overview on the most important measures to reduce
ammonia emissions from buildings shows that there is a variety of
methods available which, however, are not sufficiently investigated, yet.

It is obvious that ammonia emissions must be considered losses of
nitrogen which is a valuable fertilizer. Therefore, it is necessary to minimize ammonia emissions not only for environmental
reasons but also from an economical point of view. Costs for
reducing ammonia losses should be taken into account. However,
one must be aware of preventing other environmental problems to
occur like nitrate leaching or N_2O emissions while reducing
ammonia emissions.

8. Conclusions

1. Ammonia emissions from animal husbandry are part of the N-cycle in agriculture. 37 % of all N-losses are losses in form of ammonia.
2. Animal production has a nitrogen efficiency of 17 % only.
3. Ammonia emissions from buildings and stores comprise about 1/3 of the total emission of ammonia from animal husbandry; about 2/3 are deriving from application.
4. There are distinct differences among the animal species in respect to the emission rate per livestock unit (LU): Cattle moderate, pig and poultry high.

Measures to prevent ammonia losses from buildings are:

1. Low pH-value of the manure; lowering by acids.
2. Low temperatures diminish manure decomposition.
3. Bedding can decrease ammonia release if sufficiently used.
4. Removal of the urine from the building is more important than the feces.
5. Drying (ventilation) of poultry manure if applicable.
6. Restricted N-feeding.
7. Feed and slurry additives; its effect is still uncertain.
8. General hygienic rules (e.g. clean and dry pens, ventilation system according to ventilation norms, e.g. DIN-norm 18910, relative humidity 60 to 80 %, animal density not less than 3 m³/fattening pig).

Future requirements:

There is an urgent need for research
- on animal keeping systems with low emission levels
- on the possibilities to reduce N-excretion by N-restrictive feeding
- on the prospects of feed and slurry additives (there is a lack of standardized test and evaluation procedures for these commercially offered products)
- Further study is required on the ammonia related processes in animal manure

All these measures and efforts should not turn away from the question wether it is necessary to maintain and support the present form of intensive animal production which is likely to pollute air and water and contributes to the damage of forests and ecosystems. This should also be considered in relation to the present situation of overproduction on a large scale in animal production in Europe.

This comment is not only related to the emission of ammonia but on the emission of other gases like methan, nitrogen oxide and carbon dioxide which are considered to contribute to mechanisms related to the climate change problem.

9. References

(1)   Malanchuk, J.L. and J. Nilsson (1989). The role of nitrogen in the acidification of soils and surface waters. Miljorapport 10. Nordic Council of Ministers. DK-Kobenhaven.

(2) Isermann, K. (1990). Die Stickstoff- und Phosphoreinträge in
     die Oberflächengewässer der Bundesrepublik Deutschland durch
     verschiedene Wirtschaftsbereiche unter besonderer Berück-
     sichtigung der Stickstoff- und Phosphor-Bilanz der Landwirt-
     schaft und der Humanernährung. DLG-Forschungsberichte zur
     Tierernährung (in the press).
(3) Hartung, J. (1988). Tentative calculations of gaseous
     emissions from pig houses by way of the exhaust air.
     In: Nielsen, V.C. et al. (eds.): Volatile emissions from
     livestock farming and sewage operations.
     Elsevier Appl. Sci. Publ., London and New York.
(4) Anonymus (1989a). Emissionen von Ammoniak - Quellen - Ver-
     bleib - Wirkungen - Schutzmaßnahmen. Arbeitsmaterialien des
     Bundesamtes für Ernährung und Forstwirtschaft Frankfurt/M.
(5) Buijsman, E., H.F.M. Maas and W.A.H. Asman (1987).
     Anthropogenic NH³ emissions in Europe.
     Atmos. Environm. 21, 1009-1022.
(6) Anonymus (1989b). Plan van aanpak beperking ammoniak-emissie
     van de landbouw. Vz. 892905. Minister van Landbouw en
     Visserij and van Volkshuisvesting, Ruimtelijke Ordening en
     Milieubeheer, NL-'s-Gravenhage.
(7) Burton, D.L. and E.G. Beauchamp (1986). Nitrogen losses from
     swine housings. Agric. Wastes 15, 59-74.
(8) Miner, J.R. (1974). Odors from confined livestock pro-
     duction. Environ. Protection Technol. Ser. EPA-660/2-74-023
     U.S. Environmental Protection Agency, Washington, D.C.20460.
(9) Frenken, A. (1989). Stickstoffverluste aus verschiedenen
     Stickstoffverbindungen des Legehennenkotes während der
     Lagerung in unterschiedlichen Haltungssystemen. Bonn,
     Rheinische Friedrich-Wilhelms-Univers., Diss. agr. (Thesis).
(10) Kroodsma, W., R. Scholtens and J. Huis (1988). Ammonia
     emission from poultry housing systems. In: Nielsen, V.C. et
     al. (eds.): Volatile emissions from livestock farming and
     sewage operations. Elsevier Appl. Sci., London and New York.
(11) Kellems, R.O., J.R. Miner and D.C. Church (1979). Effect of
     ration, waste composition and length of storage on the
     volatilization of ammonia, hydrogen sulfide and odors from
     cattle waste. J. Anim. Sci. 48, 436-445.
(12) Kowalewsky, H.-H. (1981). Messen und Bewerten von Geruchs-
     immissionen. KTBL-Schrift 260, Darmstadt.
(13) Henkel, H. (1989). Über die Minimierung der N und P Aus-
     scheidungen beim Schwein. Christian-Albrechts-Universität
     Kiel. Paper collection.
(14) Franz, P., D. Dreyer, F.-J. Romberg and A. Salewski (1989).
     Mit angepaßter Rohprotein- und Aminosäurenversorgung die
     N-Ausscheidungen bei Mastschweinen vermindern. Schweine-
     Zucht und Schweine-Mast 37, 400-402.
(15) Goodall, S.R., S.E. Curtis and J.M. McFarlane (1988).
     Reducing aerial ammonia by adding Micro Aid to poultry
     diets: Laboratory and on farm evaluations. Proc. 3rd Int.
     Symp. Livestock Environment, Toronto 25-27 Apr. 286-290.
(16) Reece, F.N., B.J. Bates and B.D. Lott (1979). Ammonia
     control in broiler houses. Poultry Sci. 58, 754-755.
(17) Carlile, F.S. (1984). Ammonia in poultry houses: A
     literature review. Worlds Poultry Sci. 40, 99-113.
(18) Malone, G. W. (1987). Chemical litter treatments to control
     ammonia. National Meeting Poultry Health and Condemnations,
     Ocean City, New Jersey, Oct. 1987.
  - DIN-norm 18910 (1990). Climate in Animal Houses.
     Beuth Verlag, Berlin.

AMMONIA EMISSION FROM DAIRY AND PIG HOUSING SYSTEMS

J. OOSTHOEK, W. KROODSMA & P. HOEKSMA

Institute of Agricultural Engineering (IMAG),
P.O. Box 43, NL-6700 AA Wageningen, the Netherlands

Summary

This paper describes current research into three aspects of NH_3 emissions from housing systems for dairy cows and fattening pigs.

For dairy cows the emission has been measured from cubicle houses with slurry cellar below the slatted floor. The emissions were measured in the housing season and later when the cows were housed only at night. Emissions were measured from the cellar, from slatted floors and from concrete floors. Floor washing was compared with floor scraping.

For fattening pigs the emission was measured from three slurry systems: full and partial underfloor storage and separation of faeces and urine under the slatted floor on a sloping floor with outdoor storage, with the faeces being removed twice a day by means of a scraper. The relation was studied between emission from indoor slurry storage and the surface area of slurry, as well as the effect of the actual level of slurry.

Connected to the above, research was carried out into emission–reducing effects of flushing the area below the slatted floors in pig houses. The systems dealt with are: collecting and removal of faeces and urine in a liquid by two related systems, and flushing faeces from a sloping floor. Preparation of the flushing liquid is also described.

1. INTRODUCTION

Volatilization of ammonia is a real problem in the Netherlands. This emission contributes approx. 30% of the general acidification. Legislation demands that it be reduced at short notice. Ammonia is generally found anywhere where manure and urine are exposed to air. Emission occurs from live–stock houses, slurry storage, during and after land–spreading, and during the grazing period. The research presented here is restricted to emissions from cattle and fattening pig houses.

2. AIM OF THE RESEARCH

The research aimed primarily at measuring the ammonia emissions from various housing systems under similar conditions. Additionally, information was to be collected on processes and effects relevant to the emission of ammonia.

3. DETERMINATION OF NH_3 EMISSION

To calculate the NH_3 emission, the ventilation rate was multiplied by the NH_3 concentration in the exhaust air. If necessary, a correction factor was applied for the NH_3 concentration in the outdoor air, the background concentration. The measurements were carried out in livestock houses with forced ventilation. The amount of air passing the house and the NH_3 concentration are known so that the NH_3 emission can be calculated accurately.

4. CATTLE

Cubicle houses and underfloor storage

Cattle are accommodated in tying and cubicle houses. The last few decades, however, there has been a major shift towards cubicle houses, and nowadays approx. 70% of cattle are accommodated in cubicle houses. The houses are not insulated and have natural ventilation. Measures are taken in winter to keep an indoor temperature which is approx. 5°C higher than outdoors. In case of frost, the inlet and outlet are closed in order to keep the temperature above freezing point. As there are no reliable methods to measure the ammonia emission from naturally ventilated houses, a house was provided with forced ventilation. This house can accommodate 40 dairy cows. The slurry is stored in cellars below the slatted floors and cubicles.

Measurements were made during the 1988–1989 housing season. The NH_3 concentrations were measured in the exhaust air. The measurements in the house were continued during the grazing season, when the cattle were kept inside at night. Temperatures were measured in the house so that comparisons could be made with a naturally ventilated cattle house on the same farm. The outdoor temperature was also measured.

Results

The temperature in the experimental house was 6–7°C higher than the outdoor temperature, and, compared with the naturally ventilated house, the temperature was approx. 2°C higher. It has to be stated that the naturally ventilated house has a larger air content (expressed in m^3/animal) than houses usually have in practice, which explains why its indoor temperature was often rather low. In practice one can also expect considerable variations in indoor temperatures due to the stocking density and the setting of ventilation flaps. With a view to these considerations, the temperatures in the experimental house may be assumed to provide a reliable impression of temperatures in naturally ventilated houses. In May and June, the temperatures of the two houses were practically the same. Due to the high outdoor temperatures, there were no great differences between temperatures indoors and outdoors.

Table 1 gives the monthly data for ammonia concentration, ventilation rate, ammonia emission per hour and the ammonia emission in kg. Figure 1 shows the NH_3 emission per LSU (livestock unit) per month.

Items/Month	January	February	March	April	May	June
Concentration (mg/m^3)	4.77	4.66	3.91	4.03	4.08	3.04
Ventilation (m^3/h)	11 116	12 518	15 460	14 957	20 102	19 533
NH_3 emission (g/h)	53.3	57.7	60.1	59.4	80.9	65.1
NH_3 emission (kg/month)	38.5	38.6	43.8	43.8	60.2	46.8

Table 1: Monthly NH_3 emission figures for a cattle house for 40 cows

33

NH3
(kg/LSU/month)

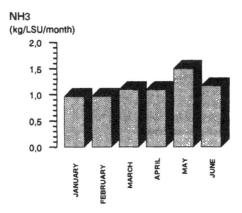

Figure 1:   NH₃ emission per LSU (livestock unit) per month

It appears from Table 1 and Figure 1 that the emission values from January till April are on an even level. There is a tendency, though, for the emission to increase in spring. The higher outdoor temperatures cause the ventilation rate to increase. This is especially expressed in the summer months. Despite the fact that the cows are in the house only at night, the emission is higher than in the actual housing period. During the day the emission falls because the slatted floors dry up and no new urine is added, but this does not alter the fact that the emission remains rather high and even increases as soon as the cows enter the house again. This is clearly shown in Figures 2 and 3. Figure 2 shows the average emission over the day for the period from 1 till 18 May, when the cows were indoors day and night, whereas Figure 3 shows the average values for the period from 19 till 31 May, when they were in the house only at night.

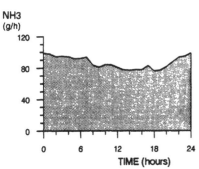

Figure 2: Average 24-h emission, with the cows being indoors day and night (1-18 May)

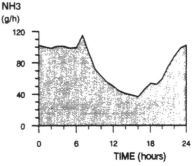

Figure 3: Average 24-h emission, with the cows being indoors only at night (19-31 May)

Figure 3 gives a view of the fluctuation in emission in the grazing period. As soon as the cows leave the house, the emission falls, till they enter the house for the night after being milked. Once they are indoors, the emission rises again as fresh urine moistens the slatted floors and the surface of the slurry. There is a remarkable peak between 6 and 7 a.m. This is the time when the cows are taken from the cubicles to be marshalled in the collecting yard before the milking parlour. Within a few minutes' time, there is much excretion of faeces and urine.

Other housing systems

In addition to cubicle houses with slatted floors and slurry cellar, there are cubicle houses with a concrete floor from where a dung scraper takes the slurry to outdoor storage.

All houses being naturally ventilated, measuring is a problem. For that reason, a method has been developed for comparative measurements. The method makes use of a Lindvall sampling box, with open bottom. It has controllable fans on two opposite sides. It is placed on a surface to be measured (area covered: 1 m²). By sending an airflow past the surface and measuring the NH_3 concentrations in the incoming and outgoing air, an impression can be obtained of the amount of ammonia which is emitted from this surface.

Results

The measurements were carried out four times for different situations, for one hour each time. The average emission values per situation have been given in Figure 4.

A: Slatted floors soiled
B: Slatted floors scraped clean
C: Surface of slurry
D: Total of A & C
E: Concrete floor soiled
F: Concrete floor scraped clean
G: Concrete floor washed

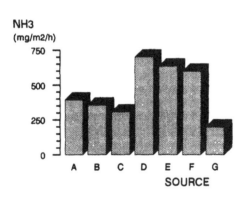

Figure 4:   Ammonia emission in mg $NH_3/m^2.h$

The results show that:
–   The total emission from a soiled slatted floor and the surface of the slurry is approximately the same level as the emission from a soiled concrete floor.
–   Scraping the slatted and concrete floors hardly reduces the ammonia emission.
–   The greatest reduction in ammonia emission is caused by washing the concrete floor.

5. PIG HOUSES

The research into the ammonia emission from fattening pig houses dealt with two items:

- The effects of three housing systems with different principles for collecting and removing slurry have been examined. This was done at the Research Station for Pig Husbandry in Rosmalen.
- The effects of three systems for flushing or collecting faeces and urine in liquid are examined. The research is being performed at the experimental farm of Sterksel.

Research into housing systems

The three housing systems involved in the research project are shown in Figure 5 and can be described as follows:

- House A: The house has a partly slatted floor, the underfloor area being fully occupied by the slurry cellar. The cellar is approx. 1.00 m deep and has a capacity for more than six months.
- House B: The house has a partly slatted floor, with slurry being stored in the channels below the slatted floor only and being removed once a week.
- House C: The house has a partly slatted floor. The cellar below the slatted floor is sloping. As a result, urine runs off immediately to be stored outdoors. A dung scraper removes the faeces twice daily.

Results

The research was carried out in 1988. The measurements were performed for one year, which were three successive fattening periods. Three values were included: the NH_3 concentration in the exhaust air, the ventilation rate, and the indoor temperature. The NH_3 emission was calculated from the average ammonia concentration and the ventilation rate. Table 2 gives the values found for each of the three periods. The amount of pigs per fattening period was 96.

	Unit	Housing system		
		A	B	C
Fattening period 1				
Temperature	°C	20.1	20.1	20.0
Emission per pig place	kg	0.90	1.19	1.29
Fattening period 2				
Temperature	°C	23.1	21.8	19.9
Emission per pig place	kg	1.21	1.12	1.26
Fattening period 3				
Temperature	°C	20.0	18.2	15.8
Emission per pig place	kg	0.85	0.67	0.52

Table 2: Emission and temperature data for three house types

For period No. 1 the temperature is the same for the three housing systems. Housing system C gives the highest emission.

For period No. 2 the temperature in housing system C is the lowest. Especially in this system, relatively much dunging took place outside the slatted floor area. This may have contributed to the NH$_3$ emission.

Period No. 3 scores the lowest emission values. The indoor temperatures are a few degrees lower than in the two previous fattening periods. For housing system C the temperature is even lower than in normal practice. The lowest emission figure has been calculated for house C. In this period the lying area was not soiled.

The housing systems and emission values per fattening place per year are presented in Table 3.

House A

House B

House C

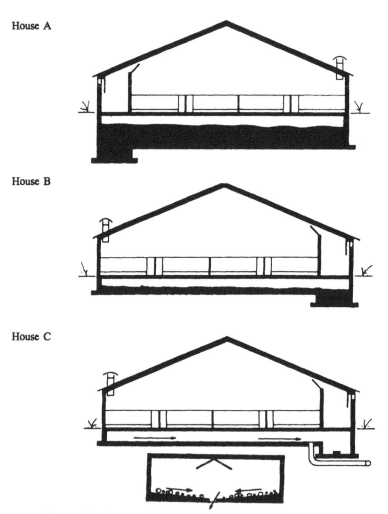

Figure 5: The housing systems involved in the research

The results show that there are no substantial differences in the total emission over a year. The temperature conditions are not always comparable. The emission from house C, where no urine and faeces are stored indoors, is not lower, despite the lower temperature.

A few conclusions are:

- The large surface area of slurry in house A with full underfloor storage does not affect the emission.
- The frequent removal of slurry from house B (once weekly) does not reduce the emission.
- The direct run-off of urine and the twice daily removal of faeces from house C do not on average show a reduction in NH_3 emission compared with the usual systems.

The results show that the quick removal of manure, a low amount of slurry stored in the house, and even the immediate separation of urine fail to provide solutions to the problem of NH_3 emissions from fattening pig houses.

Housing type	Depth of storage (m)	Frequency of removal		Kg NH_3/ pig/year
Full underfloor storage	1.0	2–3x/year		3.0
Storage channels below slatted floors	0.4	1x/week		3.0
Underfloor separation of faeces and urine	0.4	Urine: Faeces:	continuously 2x/day	3.1

Table 3: NH_3 emission per fattening place per year

Research into the flushing of pig houses

The research into the effects of flushing pig house floors was carried out at the experimental farm of Sterksel, in cooperation with the Research Station for Pig Husbandry (PV). The measurements were and are being performed in 1989 and 1990. For the flushing experiments, three fattening pig houses with partial underfloor slurry storage were arranged. The systems are described below and shown in Figure 6.

- House 1: A layer of 10 cm of flushing liquid is always present below the slatted floors to collect faeces and urine. At regular intervals, the mixture of flushing liquid, faeces and urine is replaced. Valves in the cellar floor are opened, the mixture is drained off through tubes, and a new layer of flushing liquid is applied.
- House 2: The principle of house 2 is similar to that of house 1, except for the removal of the mixture and the provision of new flushing liquid, which are performed by means of a pump.
- House 3: There is a sloping floor below the slatted floors. A few times daily, faeces and urine are flushed.

The systems are compared to fully slatted floors with prolonged storage underneath. A second reference is a pig house with partly slatted floors from where the slurry is removed every other week.

The flushing liquid

The required flushing liquid is prepared by means of solid–liquid separation of the mixture of faeces, urine and flushing liquid obtained from the pig house. The solids (11–15% dry matter) are disposed of as manure. The liquid (approx. 2% dry matter) is aerated to convert ammonia into nitrate (nitrification) followed by sedimentation. A part of the sludge is returned to the aeration tank.

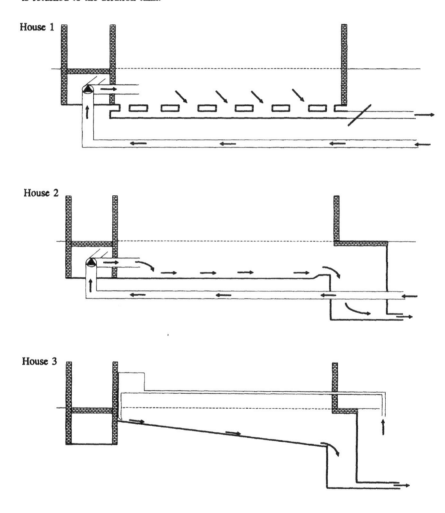

Figure 6: The systems in which flushing is investigated

Now, the flushing liquid is ready to be re-used. Its NH_4^+ concentration is about 0%, whereas its NO_3 concentration is low. When faeces are collected in the flushing liquid, a demand for oxygen occurs, for the manure contains (easily) degradable organic matter. The nitrate is then converted into nitrogen which escapes (de-nitrification). Consequently, the flushing liquid contains very low quantities of mineral nitrogen.

The diagram in Figure 7 shows how the flushing liquid is made.

Results

The NH_3 emission is calculated from the NH_3 concentration in the exhaust air and the ventilation rate, which are measured. The project not being concluded yet, only provisional results can be given.

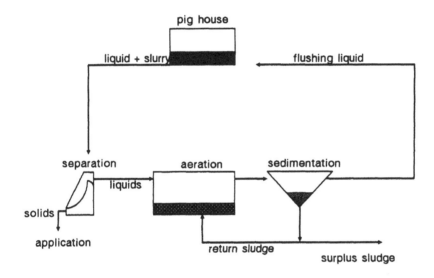

Figure 7: Process of flushing liquid treatment

The flushing systems are compared with the reference pig houses. Figure 8 gives the cumulative NH_3 emission values for 200 days for the systems.

The lowest emission values are achieved by the system whereby faeces and urine are collected in a layer of flushing liquid below the slatted floors. The best results were obtained for the system where the flushing liquid and the slurry are provided and removed by drainage (House 1). The difference with replacement by pump increases in the course of time.

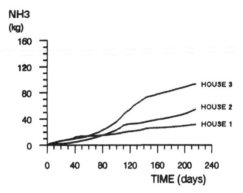

Figure 8: Cumulative NH₃ emission from houses equipped with flushing

Higher emission values are found by the system used in house No. 3 where faeces and urine are flushed with liquid. This is caused by the repeated discharge of urine on the sloping floor between flushing times.

Figure 9 shows the cumulative NH₃ emissions from the two reference systems.

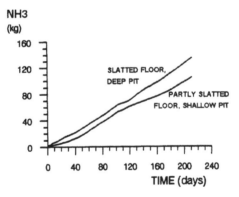

Figure 9: Cumulative NH₃ emission from reference systems

The highest emission is measured from the slatted floor with prolonged slurry storage. This is not surprising as the emitting surface is much larger. The results for the three flushing systems and two references are shown in the diagram of Figure 10.

The diagram shows that the emission from the partly slatted floor is only approx. 20% less than that from the fully slatted floor. The emission reduction achieved in the two best flushing systems is between 60% and 70%; these are the two systems with a layer of flushing liquid below the slatted floors.

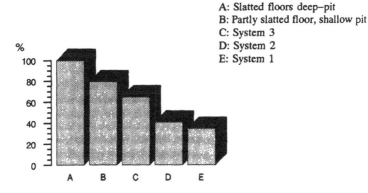

A: Slatted floors deep–pit
B: Partly slatted floor, shallow pit
C: System 3
D: System 2
E: System 1

Figure 10: Relative ammonia emissions from five systems

6. CONCLUSIONS

It appears that approx. 1 kg NH_3 month cow is emitted during the housing period. In the grazing season, despite the fact that the cows are housed only at night, the emission tends to be higher. Ammonia emission not only occurs from the storage, but also from the slatted floors, and, to about the same degree, from concrete floors. Washing of floors appears to have a much greater emission–reducing effect than scraping.

NH_3 emissions from three pig houses with different slurry systems (three fattening periods per year) were 3.0 kg NH_3 year pig place for houses with underfloor storage, and 3.1 kg NH_3 year pig place for systems separating faeces and urine. Ammonia emission from pig houses with indoor slurry storage is not only related to the surface area of slurry. Emissions occur from both slatted floors and soiled lying areas. A high or low level of the slurry does not affect the NH_3 emission. Much emission occurs from the system separating faeces and urine, despite the immediate run–off of urine and the frequent removal of faeces. With this system, dunging took place also outside the slatted floor area.

Research is continued into the effect of flushing the floor of pig houses below the slatted floors. Faeces and urine are collected in a shallow layer of flushing liquid or flushed over a sloping floor. The flushing liquid is the processed liquid fraction of the slurry. The systems collecting faeces and urine in a shallow layer of flushing liquid, compared with a reference system (a fully slatted floor with prolonged underfloor slurry storage) appears to be the best flushing system with a reduction in ammonia emission by approx. 70%.

7. RECOMMENDATIONS
- The measurements with the Lindvall box in cattle houses have shown that washing the floor results in a reduction in NH_3 emission. The washing of cattle house floors has to be investigated further.
- The measurements in fattening pig houses have shown that a few flushing systems can strongly reduce the NH_3 emission. These systems work with a treated liquid obtained from the slurry. Optimization of the systems investigated deserves more attention.
- For pig houses, research into the residual emission is advisable; the lying area and the slatted floor have to be taken into account.

COMPARISON OF THE EFFECTS OF DIFFERENT SYSTEMS ON AMMONIA EMISSIONS

H. MANNEBECK and J. OLDENBURG
Institute of Agricultural Engineering, University of Kiel
Olshausenstr. 40, D - 2300 KIEL, West Germany

Summary

On more than 200 practical farms odour and ammonia emissions from different species of amimals and different housing systems have been measured under different weather conditions. The influence of ambient air conditions, animal activity, animal species so as cattle, pigs and poultry and different housing and keeping systems as well as different manure systems have been investigated.

1. Introduction

Up until the present time ammonia was only taken into consideration in relationship to animal perfomance capacity and the health of the stall personnell dependent on the concentration of ammonia present in the stall air.

The ammonia emission expected predicted from domestic animals has now been brought into connection with the increasing earth acidity and dying forests. (5).

This theory is substansiated by the fact the substantial amounts of ammonium has been determinded in rain water in the vicinity of animal accomodation (2).

Different measures of influence of odour and ammonia emissions have been calculated and compared with emissions from cattle, pigs and poultry during systematic tests carried out by the Institute of Agricultural Engineering of the University of Kiel, West Germany.

The emissions are based in the following on a livestock unit (LU = 500 kg live mass) and on time. The ammonia emissions are given in gramm ammonia per hour and LU (g NH_3 h^{-1} LU^{-1}).

The ammonia concentrations were determined in individual momentary measurements using DRÄGER-Capillary tubes and in long term measurements on the basis of the conductivity of a test reactance.

The basic ingredient is according to the animal in question, its age and type of foodstuffs used in the resultant urine content in the liquid manures and dung which dependent on the pH-value and the temperature of bacteria is converted to nitrogenous ammonia is present as volatile ammonia, which is in turn to dependent on the ambient temperature dissipated in the environment (1).

2. Influence of Air Conditions and Animal Activity on Ammonia Emissions

When animals are to be held in stalls then an optimal temperature suited to the animals in question is to be aimed at.

The necessary air volume required to despense with the resultant warmth, steam and harmful gases is determined by the energy and saturation deficit of "fresh air" as opposed to the "stall air". This increasing fresh air temperatures lead under otherwise identical conditions to increasing air volume rates. Further deciding factor is the heat generated in the stall which is in the main in direct relationship to the activity of the animals involved.

Both the air conditions (day-night variations, weather influences and season) and the degree of animal activity (feeding time, rest periods, light influence) change and repeat themselves regularly according to a cyclus.

Furthermore changing temperatures in the stalls which are unavoidable with normal ventilation systems lead to different conversion and exhaust in odour and ammonia emitting organic substances.

The influence of ventilation air conditions and animal activity during daily processes have been clearly identified by investigations made each on a battery-hens stall and on pig fattening houses (figure 1).

The higher ventilation air temperatures normally occur simultaneously with the periods of higher animal activity due to the fact that the stalls are brightly illuminated and because the feeding times also occur in this period. The ammonia emissions increase steeply from dawn onwards analog with the ventilation air temperature and animal activity. The ammonia emissions reach their maximum some hours before the temperature zenith. Due to the fact that the conversion of the organic substance as the basis for the creation of ammonia does not react correspondingly fast to temperature variations only limited ammonia emission increases occur with increases in temperature.

The strongest day/night variations occur in the poultry battery. The laying hens are illuminated not by daylight but by an illumination program. The hens have absolute quietness at night and none at all during the day. They differ so from pig fattening pens in which the animal activity is not affected by such an extreme by the day/night exchange.

The difference in the NH_3-emission in pre- and final pig fattening pens is also to be explained by the differences in animal activity. During the pre-fattening period feeding is carried out "ad libitum" and in the final fattening period feeding is rationed. The animals are more active at feeding times. During the final fattening

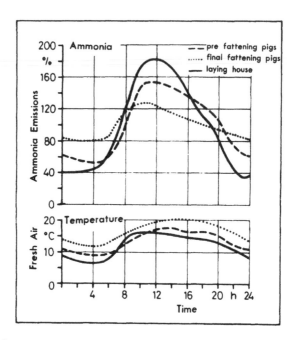

Figure 1:
Standardised ammonia emissions of a battery hen stall and two pig fattening houses over a period of 24 hours and corresponding ventilation air temperatures.

period the feeding times have a 12 hour period and the increased animal activity is distributed over the whole day whereas the "ad libitum" fed animals eat according to their moods during the second half of the day.

In spite of the stall system conditioned differences the stalls investigated have shown in part higher ammonia emissions during daytime than at nights. This phenomena must be taken into account in stall investigations which are mainly carried out during the daytime. Long term temperature variations effect both the conversion of the organic substance and thus on the release of ammonia and also on the level of the air-rate as a whole (figure 2).

In the investigated tying stall for cattle the total level of ammonia emission with clearly less than 0,5 g NH_3 h^{-1} LU^{-1} was very low. The increase due to increasing ventilation air temperature was also very small. These very low NH_3 emissions from cattle houses can be traces to two separate factors:

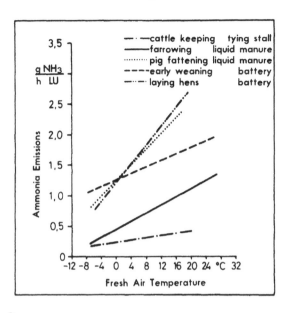

Figure 2:
     Ammonia emissions of diverse stall systems in dependence of the day period on ventilation air temperature.

- the proportion of emissions active area of the total area is comparatively small in the cattle pens because only approximately 20 % of the pen area is allocated to the dung channel.
- The cattle-pens are usually cleaned twice daily. Due to this the normal conversion process is retarded because of the very short time intervals involved.

3. Ammonia Emissions from Different Animals and Different Housing Systems
     The ammonia emissions are the higher the longer the amount of time the animal droppings remain in the pen and the more (old) organic mass is lying in the air current. This double effect leads to the high NH_3-emissions with ground kept poultry (figure 3).
     The higher NH_3-emission of littered pig fattening pens as opposed to liquid manure systems can thereby also be explained. The fact that littered pig-pens emitt more ammonia as strawless ones, in those where the laying time of the liquid manure is often longer indicates that the gas exchange between the pen area and the

liquid manure surface under the slatted flooring is relatively small. The gas exchange when using totally slatted flooring as opposed to a partially slatted floor is greater. The liquid manure surface in a totally slatted floor is greater. The liquid manure surface in a totally slatted floor pen corresponds to the basal surface of the pen.

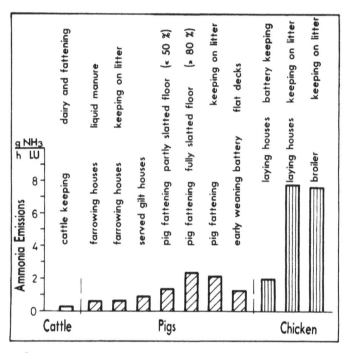

Figure 3:
        Average ammonia emissions of different stall-systems on the basis of the yearly average temperature.

        With battery hens which normally only produce approximately 1/4 the ammonia emissions similar high emission rates are to be expected, when the droppings remain lying in the stall for several months such as in the case when keeping poultry on litter.

4. Emissions Arising from Broiler Keeping
        Broiler farms may be used as an example to indicate the specialities of IN/OUT-systems with respect to odour and ammonia emissions (figure 4).

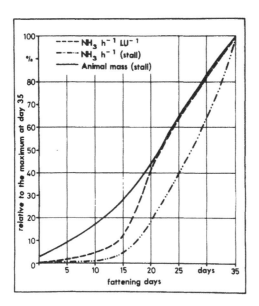

Figure 4:
   Ammonia-emissions from a broiler farm with an increasing fattening period.

   Broiler fattening farms use the IN/OUT-strategy because of reasons of hygiene and because of work saving reasons: The size of the animal mass increases enormously over the fattening period. This leads to an increase of 30 : 1 over a period of 35 fattening days. As the production period is governed by a fixed end weight only a few weeks are needed. In this case it is usual that the pens be littered with straw at the beginning of the fattening period and first cleaned out on conclusion of the same. The original straw layer is odourless and emitts no ammonia at all. This layer of straw receives a great deal of chicken droppings over the fattening period which in the high relative temperature of the chicken battery quickly converts.
   During the first 10 days there is relatively little droppings in the pen and therefore correspondingly little ammonia emissions. First conversions begin to occur. After approximately 15 days the ammonia emissions have risen to a level of approx. 2 to 3 g NH_3 h^{-1} LU^{-1} which is the level normally associated with other stall systems (see figure 3). Thereafter both the emissions related to the animal mass itself increase as exponential multiplicators.

An emission value of approx. 30 g NH_3 h^{-1} LU^{-1} was determined on day 35. Odour emissions develope in a similar manner. At the end of the second week the odour emissions were determined as 25 OU s^{-1} LU^{-1} and at day 35 determined as being 225 OU s^{-1} LU^{-1}.

The stalls are emptied of animals on day 36 and cleaned. The ventilation is switched off and the stall emissions are once again zero. Such a produce leads to a permanent cycle of high emission periods interspersed with periods of zero emission (figure 5).

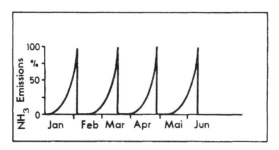

Figure 5: Ammonia emissions from a broiler farm the year round

5. Methods to Decrease Emissions

Emissions can be minimised by using the appropriate stall systems and the corresponding stall management methods. Unwanted gases which have already formed must be rendered harmless by an exhaust air cleaning system. Should one wish to reduce emissions than one must start with the organic material:
- the stall systems should have a small emission profile
- bio-chemical conversions require water. Conversions do not occur in a dry environment
- Use of a ventilation system which neglects the space between the liquid manure surface and the perforated flooring reduces emissions as the transport of emission substances is then avoided.

A conflict of interest can occur here at the underfloor ventilation system used to suck off harmful gases below the slatted floor leads to transport of emission substances and thus to increased emission.
- A cooled air ventilation leads to smaller air rates and/or lower stall temperatures with the corresponding effect on emissions. Use

of an earth heat exchanger (4) is one method with which emission peaks can be stopped.

In future emission behaviour must be observed when developing new stall systems. Separation of the functional areas: feeding, laying and defecating gives larger free-spaces for ventilation an concipation of the stall flooring.

Ventilation exhaust purification units can also be used (6).

Production procedures with large variations in animal mass such as IN/OUT-systems or with pig and poultry fattening systems the problems described will be partially agravated and increased due to the enormous variation in animal mass. The ventilation system must be able to cope also with the absolute emission peak; i. e. appropriate configuration (costs) although this requirement is only for a few weeks in the year.

REFERENCES

(1) ELLIOT, H. A.; N. E. COLLINS (1982): Factors affecting ammonia release in broiler houses, Transactions of the ASAE, S. 413

(2) HARTUNG, J. (1986): Zum Ammoniumgehalt des Regenwassers in der Umgebung eines Schweinemaststalles, Staub - Reinhaltung der Luft (46), H. 10, S. 429-431

(3) OLDENBURG, J. (1989): Geruchs- und Ammoniakemissionen aus der Tierhaltung, KTBL-Schrift 333, Darmstadt

(4) TIEDEMANN, H. (1987): Besseres Stallklima durch Erdwärmetauscher, Landtechnik (42) H. 9, S. 364

(5) VAN DER EERDEN, L. J.; H. HARSSEMA und J. V. KLARENBEEK (1981): Stallucht en planten. De reatie tussen bedrifsomvang en den kauss op beschadigung van gewassen random intensieve verhouderijbedrijven, IPO-Rapport R-254, Wageningen (NL)

(6) WÄCHTER, G; J. JANSSEN (1977): Behandeln der Abluft aus Tierhaltungsbetrieben zur Senkung von Geruchsstoffimmissionen, Grundlagen der Landtechnik (27) H. 3, S. 88

AMMONIA EMISSION FROM TWO POULTRY MANURE DRYING SYSTEMS

L. VALLI, S. PICCININI and G. BONAZZI

Centro Ricerche Produzioni Animali
Via Crispi 3 - 42100 Reggio Emilia - Italy

Summary

Tests were carried out at two poultry farms for laying hens with two different operational systems for the partial drying of poultry manure on the farm itself: the first uses a composting plant and the second "ventilated deep pit" housing. The aim was to assess the characteristics of the different systems, particularly the ammonia emissions.
Specific ammonia emissions per bird proved extremely high in the case of the composting plant (0.951-1.628 g/hen d) whilst they were considerably lower in the ventilated deep pit plant (0.176 g/hen d).

1. INTRODUCTION

For some time, plant proposals have been developed which use different technical solutions in order to treat the droppings from the laying hens in such a way as to obtain a stabilized product with a high dry matter content which is easy to store, transport and distribute in the field.
Two technological solutions in particular have come to the fore in Italian farms: composting plants and ventilated deep-pit hen-houses.
In the first the poultry manure, extracted daily with a high moisture content (80%) from the pits under the cages, is mixed with lignocellulose residue (staw and/or broiler litter in particular) or with some quantities of recycled pre-dried poultry manure, until a moisture content of approximately 70% is obtained. The mixture then goes on to the composting stage which takes place in horizontal pit or trench reactors 60 to 100 metres long. Here the product is stirred each day by a special machine and is affected by an intense and spontaneous aerobic and esothermic-type fermentation which causes progressive dehydration of the product. In order to speed up this drying process, the reactor is enclosed in a hothouse shed thereby increasing the temperature of the air coming into contact with the product. The plant is capable of reducing the moisture of the mixture, with a retention time of 30-50 days, from 65-70% to 25-30%, especially during the summer season.
In the second plant the poultry manure may be stored under the cages of the laying-hens without being moved for a whole productive cycle. The building consists of two tiers: the lower tier functions as a store for the droppings whilst the upper

tier houses the hens. Air-conditioning of the premises and drying of the manure are made possible by a forced ventilation system.

Two farm-scale plants were monitored in order to analyse the results from the two different drying systems in terms of the characteristics of the manure obtained, the impact on the environment and in particular the ammonia emissions.

2. POULTRY MANURE DRYING IN DEEP PIT PLANTS

In farms equipped with a ventilated deep pit storing and drying system for manure, dehydration of the poultry droppings takes place using both the metabolic heat produced by the animals and the heat which forms during the fermentation of the manure itself. The building that houses the hens is composed of two interconnected tiers: the lower tier is the place where the manure is deposited, dried and stored whilst the upper tier is the actual breeding area which houses the cages of the laying hens.

The premises are ventilated by means of fans which extract the air from the manure storing compartment, withdrawing it from the upper tier and making it follow a C-shaped course (fig.1) during which it heats up whilst skimming the roof of the building, then crossing through the cages and absorbing the heat produced by the animals (1). During a productive cycle the manure in the heaps tends to dry out and stabilize itself, thanks to the strong ventilation and the esothermic fermentation which starts in it.

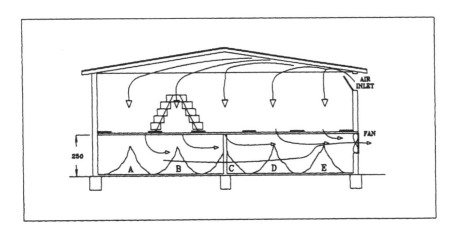

Fig.1: The C-shaped ventilation diagram of the deep pit poultry house

2.1. The tests carried out

The breeding building where the tests take place is 12 m wide and 120 m long, and house 50,000 laying hens arranged in 5 rows of 5 tiers high cages. The C-shaped ventilation described above is brought about using 15 extractor fans 30,000 m^3/h each. There are 4 adjustable levels controlled by a thermostat: 8 extractor fans are active at the first stage, 11 at the second, 13 at the third and 15 at the fourth.

The aim of the control programme was to assess, on the one side, transformations which occur in the manure during the storing period as well as its characteristics when finally removed and, on the other side, the ammonia emmissions with ventilation.

The monitoring campaign started four months after the beginning of the breeding cycle and continued to the end of it. Controls were made at monthly intervals. In 3 of the 5 heaps (called A/C/E) manure samples were taken and the following analytical determinations were made:
- total solids (TS);
- volatile solids (VS);
- total Kjeldhal nitrogen (TKN);
- ammonium nitrogen (N-NH$^+_4$);
- total phosphorous (Pt);

At the same time and in the same heaps temperatures were also measured in 5 different positions lengthwise and at 3/4 different depths (30/60/100/130 cm from the surface layer).

Owing to problems with the sampling equipment it was not possible to take samples to determine the ammonia concentration in the exhaust air during the same breeding cycle, only in the subsequent one. Samples were taken at fortnightly intervals for a period of 5 months, starting from the 7th month of the cycle, during the period which corresponds to the transition from the maximum fermentation activity phase to the subsequent stabilizing one. The samples were taken directly from the mouth of two of the active ventilators in two positions. The concentration of NH_3 in the air was determined by bubbling through an H_2SO_4 trap and subsequent colorimetric determination with Nessler reagent.

2.2. Results and discussion

Temperature in the heaps

The diagram in fig.2 shows average temperatures for each of the three heaps and for the whole test period.

During the first 7 months the temperatures maintain an average above 40°C and, in the case of heap A, which is the farthest from the ventilators, they come very close to 50°C. Such thermic levels demonstrate how predominantly esothermic-type fermentations take place in the heaps, due to the aerobic transformation of the organic matter assisted by the ventilation to which the heaps are subjected.

The active period tends to lessen after 7/8 months and temperatures decrease to around 30-35 °C, keeping at these levels until the end of the cycle.

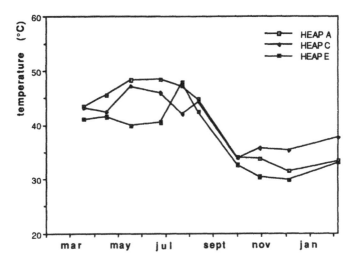

Fig.2: Average temperatures in the heaps during the drying cycle

Chemical parameters

Table 1 shows the average values of the analitical determinations on samples taken at the beginning and at the end of the cycle.

Table 1: Chemical determination on poultry manure at the starting and at the end of the cicle from ventilated deep pit house.

PARAMETERS	HEAP A		HEAP C		HEAP E	
	START	END	START	END	START	END
Total Solids (% w.b.)	28.9	50.9	27.1	57.1	29.1	71.3
Volatile Solids (%TS)	65.0	45.0	66.9	48.2	63.6	62.4
Total Nitrogen (% TS)	8.1	3.5	8.2	3.4	7.3	4.6
Ammonia Nitrogen (% TS)	4.7	1.2	6.6	0.8	4.7	0.6
Total Phosphorus (% TS)	3.3	2.8	3.3	2.5	3.2	2.0

The diagram in fig.3 shows the progression of the droppings humidity content during the breeding cycle, in the three controlled heaps. Starting from an initial value equal to about 70%, the same for all the heaps, a difference in behaviour during the following period can clearly be seen depending on the distance of the piles from the fans. Heap E is nearest to the ventilators and right from the start shows a higher level of dehydration which it maintains throughout the whole period, reaching in the summer/autumn season (which was especially hot and dry during the monitoring campaign) values less than 20%. Heaps A and C show similar trends in the dehydration process even though the final results are less positive: pile C reaches relative humidity values of just less than 45% and pile A never goes below 50%.

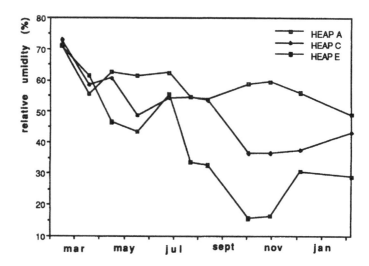

Fig.3: Relative Humidity of the poultry droppings during the drying cycle.

Even in the drying process analysed here a significant reduction in the nitrogen content during the storage phase was found (fig.4). On average, initial values of about 7-8% (d.b.) decrease to final values of about 4% (d.b.), whilst ammonia nitrogen goes down from values of 60-70% NTK to around 25% at the end of approximately a 16-month storage cycle. This reduction is doubtless attributable mainly to ammonia volatilization.

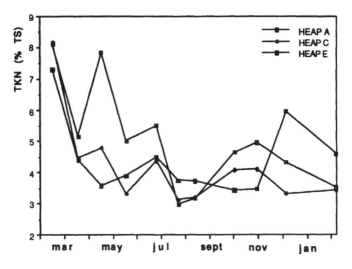

Fig.4: Total Nitrogen content of the poultry droppings during the drying cycle.

Ammonia emissions
    Table 2 shows the concentration of ammonia in the air extracted with ventilation over a 5-month storage period.

Table 2: Ammonia emission from the ventilated deep pit poultry house

SAMPLE DATE	SPECIFIC VENTILATION	NH3-N AIR CONC.	NH3-N TOTAL EMISS.	NH3-N SPEC.EMISS.
	m3/hen h	mg/m3	kg	g/hen day
11/22	4.87	2.39	516.67	0.28
12/07	2.75	2.75	136.04	0.18
01/05	2.95	2.22	227.59	0.16
01/15	2.54	1.72	52.47	0.10
01/30	2.43	3.77	164.80	0.22
02/13	2.58	2.74	118.96	0.17
02/28	3.12	1.55	86.94	0.12
03/13	3.72	1.45	84.10	0.13
AVERAGE				0.17

The concentrations are not especially high in terms of any annoying effects they could have on people, as on the whole they remain well below the TLV (Threshold Level Value) limit of 18 mg/m^3 set by the ACGIH (American Conference of Governmental Industrial Hygienists) in the United States.

3. DRYING OF POULTRY MANURE BY COMPOSTING

A composting plant made up of a rectangular pit reactor 60 m long enclosed in a hothouse shed, equipped with a stirring machine activated once a day is in operation at a farm for laying hens. The plant is loaded with a mixture of poultry manure and chopped straw at a ratio of 7:1 (by weight). A ventilator with a maximum air flow rate of 30,000 m^3/h extracts the air from the plant. The progress of this technique and the characteristics of the final product were the subject of a previous paper (2). Here it was highlighted that while the plant enables a product with good agronomical and physical properties to be obtained, the odours and nitrogen emitted through volatilization are extremely high.

With a view to reducing ammonia emissions two consecutive tests were started, with pilot-scale plants able to treat a fraction of the air extracted from the composting reactor. A washing tower with acid reagents was used in the first and a biofilter, still in operation, was used in the second.

During these tests, samples were taken to determine the ammonia content in the air extracted from the plant upstream and downstream from the treatment systems. The emissions are at their highest when the stirring machine is turning, a phase which lasts two and a half hours.

Table 3: Ammonia emission at the poultry manure composting plant

	fan working time h/y	NH3-N concentr. mg/m3	NH3-N emission kg/y
1° TEST: PILOT SCRUBBER (number of hens = 65,000)			
STIRRING MACHINE ON	1300	354±65 (n=10)	9196.72
STIRRING MACHINE OFF	7460	197±31 (n=6)	29420.75
TOTAL			38617.47
2° TEST: PILOT BIOBED (number of hens = 53,000)			
STIRRING MACHINE ON	1300	180±92 (n=11)	4688.32
STIRRING MACHINE OFF	7460	92±24 (n=11)	13699.54
TOTAL			18387.86

Tab. 3 shows the average concentration values found in the two subsequent series of tests: in the first the plant was treating manure from 65,000 laying hens equivalent to about an 11.7 ton per day load, whilst in the second the number of birds was reduced to about 53,000 with a 9.5 ton per day load.

It can be seen that ammonia concentration is in both cases extremely high reaching levels that are a source of nuisance and cause complaints from the neighbourhood. As a consequence, nowadays the plant has been stopped by the Municipal Authorities, waiting for a Regional Technical Committee advice which will extabilish the allowed emission limits.

4. TOTAL AMMONIA EMISSIONS

Evaluations of total ammonia emissions per bird were made on the basis of the amount of ammonia concentration in the air extracted from the two different monitored plants (composting plant and ventilated deep pit plant). However, quantifying the air flow rates of the extractor fans proved difficult because in both cases the ventilators are fixed on walls and have no canalization. The measures were made on each sampling date by means of a vane anemometer which gives an average value on a 6-second basis. It was positioned at various points directly at the fan discharge mouth and then the average of the values taken was worked out.

The air extraction from the hothouse enclosing the composting plant is continuous 24 hours a day all year long. The quantity of air extracted was equal to 20,000 m^3/h (±20%). The concentration of NH_3-N contained in it, as seen previously, increases in the two and a half hour working time of the stirring machine operated once a day.

Tab. 4 shows the results of calculations formulated on the basis of the data collected over the two test periods. There were specific NH_3-N emissions of 1.628 g/bird/day, corresponding to 9.04 kg/t of treated manure, in the first instance and 1.194 g/bird/day, corresponding to 5.28 kg/t of manure in the second. Presumably these differences are due to the improved straw:manure ratio made possible by the reduction of the manure load in the composting reactor.

Table 4: Ammonia specific emissions at the poultry manure composting plant

	NH3-N specific emissions		
	kg/bird y	g/bird d	kg/t manure
1° TEST	0.594	1.628	9.043
2° TEST	0.347	0.951	5.281

The values obtained don't differ considerably from the results reported by Kroodsma (3) in composting tests carried out on predried poultry manure.

In the deep pit plant, owing to the fact that the ventilation was regulated by a thermostat, it was necessary to check not only the air flow rate but also the length of time the extractor fans are in operation. An annual balance is unable to be compiled, since the tests carried out are limited to only 5 months.

An average specific ammonia emission (tab.4) of 0.176 g/bird/day was obtained. The season or the different phases of the cycle do not seem to have had obvious effects on the data collected up till now.

5. CONCLUSIONS

The evaluations that were made highlight the considerable amount of ammonia emissions that derive from the treatment of hen manure by composting and which are difficult to reduce on account of characteristics peculiar to the process (initial ammonification phase, high temperatures and high pH, low C/N, frequent stirring of the product). This is without any doubt the major drawback of the technique which causes strong uncertainties about its wide diffusion, unless it is be possible to find technically and economically acceptable solutions to treat the exhaust air.

Emissions from the ventilated deep pit plants prove considerably less thanks to the slower and less intense process which the manure is put through.

Anyway it is foreseeable that, in Italy too, poultry farmers will turn their attention to drying treatment systems which enable lower ammonia emissions, like the ventilated belt batteries systems, that are largely employed in Northern Europe (3).

REFERENCES

1) Xausa A.E. (1988) - Drying manure in a deep pit - Poultry, Oct-Nov 1988.

2) Bonazzi G. et al. (1987) - Controlling ammonia emission from poultry manure composting plants - in "Volatile emission from livestock farming and sewage operations", Proceedings of a CEC workshop in Uppsala, Sweden, 10-12 June 1987.

3) Kroodsma W. et al. (1988) - Ammonia emission from poultry housing systems.- in "Storing, handling and spreading of manure and municipal wastes", Proceedings of a C.I.G.R. Seminar in Uppsala, Sweden, 20-22 September 1988.

ODOUR AND AMMONIA EMISSIONS FROM MANURE STORAGE

M.J.C. DE BODE
Institute of Agricultural Engineering (IMAG)
P.O. Box 43
6700 AA Wageningen, the Netherlands

Summary
     In experimental slurry tanks of 4½ m³ with a diameter of 1.9 m
research has been performed into (a) the emission of odour and ammonia
during five months of storage of pig and cattle slurries, (b) the
effects of emission-reducing measures, and (c) compositional changes
of slurries during storage.
     The odour concentration measured near the tanks was low and could
be further reduced by covering the openings. The nitrogen losses
during storage caused by NH₃ volatization amounted to 5-15%. The NH₃
emission could be reduced by covering the tanks or, on the surface of
cattle manure, by crust formation. An adequately closing cover could
reduce the emission by 70-90%. Crust formation was effective only if
it had been stimulated artificially by adding chopped straw to the
slurry. In this way the emission could be reduced by 60-70%.
     Reduced emission entailed a higher ammonium concentration in the
slurry. If the share of ammonium nitrogen in the total nitrogen is increased, the manurial value of the slurry is more predictable.

INTRODUCTION
     In the past few years, there has been an increase in the number of
slurry storages in the Netherlands, and this number is expected to increase
further. This is a consequence of restrictions on the periods that manure
can be spread on the land, and of the tendency that farmers want to make a
better use of manure and so will need more storage capacity.
     Odour and ammonia are emitted from stored manure, which cause the in-
convenience of an unpleasant smell and environmental acidification due to
ammonia. To control the most serious consequences of the acidification
problem, the Dutch Government has committed itself to reduce the NH₃ emission by 50-70% of the 1980 level by the year 2000. Emissions from slurry
storages shall also be reduced. One of the means to achieve this, is
covering the storages. For cattle slurry, crust formation on the surface
can be adequate for this purpose. Crust formation can be stimulated by
adding chopped straw (1).
     When the new legislation was made, hardly any figures were known of
the emission of odour and ammonia from stored slurry and of effects of any
emission-preventing measures. In the literature, nitrogen losses had been
calculated of 4-15% on the basis of model and mass balance calculations
(2,3). On behalf of emission-reducing measures, only a few measurements
into odour emission had been made (4,5).
     In order to achieve further information, the aim of this research pro-
ject was: (a) determining the emission of odour and ammonia from slurry
tanks, (b) establishing the effects of emission-reducing measures, and (c)
investigating in-storage compositional changes of slurries.

SET-UP
     The research is carried out in experimental tanks, which are placed
next to each other. The tanks are 2 m high and have a diameter of 1.9 m.
The research project will take three years, and each year the tanks are
filled with a different type of slurry. In the first year, the tanks were

operated with pig slurry and in the second with cattle slurry. In the third
year experiments will be made with poultry slurry. A research year is split
up into a summer and a winter storage period, each of approx. five months.
This paper will be restricted to the research into the emission from stored
pig and cattle slurries.

The effects of odour and NH_3 emission-reducing measures are measured
by covering four tanks and leaving one tank without covering. The following
covering types have been selected:
- Tent: This is a superstructure, which effectively prevents rain from
  coming in. This structure allows for a minimal exchange between outside
  air and air below the covering.
- Floating foil: This covering floats on the surface, which has the advantage, if treated well, that rainwater does not get mixed with the slurry.
- Corrugated sheets: This is a superstructure too. It remains open along
  the edge of the tank, so that there can be exchange of air.
- Expanded polystyrene (tempex): This covering floats on the surface from
  where rainwater cannot be removed by pumping, so that it will dilute the
  slurry.

In addition to emission-reducing covers, the emission from cattle
slurry can be lowered as a result of crust formation. Initially, it was
attempted to let the crust form naturally. Because it took a long time for
the crust to be formed, crust formation was stimulated by adding chopped
straw. For this purpose, 4-7½ kg/m² of chopped straw was put into the tank
before it was filled with slurry.

To be able to compare the effects of the emission-reducing measures
with a situation in which no measures are taken, one or two reference tanks
are not covered. In addition to an open experimental tank, a farm-scale
tank is included in the research. By comparing the emission data for the
two uncovered tanks the practical consequences of the research are investigated.

MEASURING METHOD

The odour and ammonia emissions are measured with a large box (Lindvall box) placed over the tank plus covering or over the open tank (Figure
1). A fan provides a constant air flow of 48 m³·min⁻¹ through the box. In
an open tank, this air flow corresponds with an air speed inside the box of
1 m·s⁻¹. In tanks with a superstructure, the air speed in the box will be
slightly higher. The emission is calculated by comparing the ammonia concentrations in incoming and outgoing air multiplied by the air flow.

For the odour examination, samples are taken from the incoming and
outgoing air flows and collected in odourless teflon bags. Next, the odour
concentration is determined by an odour panel (6). Once in a month, all the
objects are sampled within half a day.

For preparing samples used for measuring the ammonia emission, parts
of the incoming and outgoing air flows are led through 85 ml of 1% phosphoric acid. The NH_3 concentration in the absorption liquid is determined
in accordance with Dutch standard NEN 6472. The emission is calculated as
follows:

$$E = ((C\ NH_{3av} * V_{av}/AV_{av})_{out} - (C\ NH_{3av} * V_{av}/AV_{av})_{in}) VS * 3600/A \qquad (1)$$

where:
E	=	NH_3 emission (mg $NH_3 \cdot m^{-2} \cdot h^{-1}$)
C NH_{3av} =		NH_3 concentration in absorption liquid (mg $NH_3 \cdot l^{-1}$)
V_{av}	=	volume of absorption liquid (ml)
AV_{av}	=	volume of air passed through absorption liquid (m³)
VS	=	volume of air over across the tank (m³·s⁻¹)
A	=	surface of tank (m²)

Picture 1: Measuring set-up

For measuring the NH_3 emission with the Lindvall box, a 24-h sampling period is required. This complicates the comparison between the objects as there tend to be considerable fluctuations in weather conditions during the measurements of individual objects. This disadvantage has to be compensated by performing a great number of measurements.

In-storage samples are taken at heights of 40 cm, 90 cm and 165, and an after-storage mixing sample is taken after completion of the period. The samples are analysed for their contents of nitrogen (Nkj), ammonium nitrogen, phosphate, dry matter, ash, pH and volatile fatty acids. The analysis results render more information on the quantity distribution of the various components in the manure. By calculating average values for the entire tank from the slurry samples taken from the three levels, a balance can be made for the various components. Mass balance calculations on the basis of Nkj make it possible also to make an estimate for the NH_3 emission. Such an estimate is less accurate than the emission measured with the Lindvall box (7).

Odour and NH_3 emissions are also measured from a farm-sized tank. In general, the same measuring method is used as for the experimental tanks. Because of the much larger size, it is not possible to cover the entire tank with a Lindvall box. Therefore, a floating Lindvall box is used, placed on floats inside the slurry tank and connected to a fan by means of an odour-free hose. The fan provides a constant air flow of 24 $m^3 \cdot min^{-1}$ (air speed = 1 $m \cdot s^{-1}$) through the box.

ANALYSES OF SLURRY

Table 1 gives the analyses of the starting material during the various storage periods.

Table 1: Composition of manure at start of storage period

	Unit	Pig slurry		Cattle slurry	
		Summer	Winter	Summer	Winter
NH₄-N	mg/l	3030	3520	3150	2490
Nkj	mg/l	4660	4900	4950	5150
P	mg/l	970	800	340	730
pH			7.7	7.4	7.6
Dry matter	%	4.2	6.8	8.1	9.2

ODOUR EMISSION

The odour concentration measured for the uncovered experimental tanks appeared to be low (Table 2), considering that the odour concentration in outside air is approx. 20 odour units (o.u.) and concentrations can occur of a few ten thousands odour units during land spreading (8). The odour from stored pig slurry was stronger than from cattle slurry, and from both, the concentration was about twice as high in summer than in winter.

Covering the experimental tanks reduced the odour concentration between 0% to 70% (Table 3), with the best reduction performance achieved in summer. The stronger reduction in summer was caused by the higher odour concentrations which usually occur in that time of year.

Table 2: Odour concentration from non-covered experimental storage tanks in odour units per cubic metre (o.u./m³)

	Odour concentration	
	Summer	Winter
Pig slurry	200	120
Cattle slurry	110	60

Interpretation of the data is complicated as covering materials emit odour themselves and differences in odour cannot be measured. The odour concentration of stored slurry being low, the analyses will easily be biased by other odours such as of covering materials. The better performance of cattle slurry may be explained by this being measured in the second year, when the covering materials had been used for more than one year and possibly emitted less odour.

Table 3: Reduction in odour emission by covering in %

	Pig slurry		Cattle slurry	
	Summer	Winter	Summer	Winter
Tent	35	15	72	42
Corrugated sheets	50	28	28	15
Floating cover	28	0	43	41
Expanded polystyrene	40	10	39	16

AMMONIA EMISSION

In these experiments the loss of nitrogen as a result of ammonia volatization was 5-15% for a storage duration of 180-250 days. This is in line with values found in the literature (2,3). Table 4 gives the values found for uncovered tanks. The emission values measured by means of the Lindvall box where clearly lower than those calculated from the slurry samples. This may be explained by the fact that the average wind speed over the tanks is 2-3 $m \cdot s^{-1}$, whereas the air speed in the Lindvall box during measurements is 1 $m \cdot s^{-1}$. As emission speed is proportional to air speed (9), the emission measured in the Lindvall box will be lower than if submitted to natural wind conditions. It has to be stated, however, that measurements with the Lindvall box are more reliable than the mass balance calculations on the basis of slurry analyses (7). The emission from farm-scale installations were also measured at an air speed of 1 $m \cdot s^{-1}$. Despite the fact that the wind speed was the same, the emission measured from the farm-scale installation was evidently higher than from the experimental tanks. An explanation cannot be given as yet.

Table 4: Ammonia emission from non-covered experimental slurry tanks in mg $NH_3 \cdot m^{-2} \cdot h^{-1}$

	Pig slurry		Cattle slurry	
	Summer	Winter	Summer	Winter
Lindvall box	600	200	300	130
Slurry samples	1800		600	
Farm-scale		650		350

There is a remarkable difference in emission speed between pig slurry and cattle slurry. From all the objects, the emission from stored pig slurry was two to three times as high than from cattle slurry, despite the fact that pH and ammonia contents of the slurries are the same. The higher emission from pig slurry is probably due to the fact that pig slurry settles and cattle slurry does not. Ammonia will more easily escape from a watery than from a viscous substance.

The NH_3 emission can properly be reduced by covering the tanks (Tables 5 and 6).

Table 5: Reduction of NH_3 emission by covering stored pig slurry in %

	Summer			Winter	
	Lindvall box		Slurry samples	Lindvall box	
	Mean	Range		Mean	Range
Tent	94	86- 97	99	84	71-96
Corrugated sheets	84	33-100	52	54	31-72
Floating cover	94	80-100	78	73	58-84
Expanded polystyrene	85	75- 93	89	78	36-87

Three of the four covering methods achieved a reduction performance of between 70 and 90%. Only the corrugated sheets scored lower, with 50% less emission. In Tables 5 and 6 the mean reduction values are given as obtained with the Lindvall box, together with the range for the individual measurements, and the results obtained from the slurry samples. Emission reduction values obtained from slurry samples are given for the summer only, because

the reliability of winter measurements is rather low. The deviations from
the mean reduction value in individual observations for the tent, floating
foil and expanded polystyrene do not exceed 20%. The potential emission
from pig slurry being higher than from cattle slurry, the former will
achieve a higher emission reduction. After all, the emission from covered
tanks appears to be independent of the type of slurry stored.

Table 6: Reduction of NH₃ emission from stored cattle slurry in %

| | Summer | | | Winter | |
| | Lindvall box | | Slurry samples | Lindvall box | |
	Mean	Range		Mean	Range
Tent	84	79-92	84	71	58-75
Corrugated sheets	50	31-61	36	46	14-79
Floating cover	86	74-94	38	82	53-96
Expanded polystyrene	81	72-93	85	78	43-94
Natural crust	37	0-65	0		
Forced crust	71	55-86	78	63	0-89

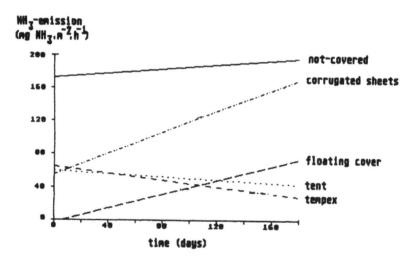

Figure 2: Curves for the emissions from stored cattle slurry in winter

The emission from cattle slurry can also be reduced by crust formation
on the surface. The crust can be formed naturally. In the experiments, this
took 1½ months, and once formed, the crust could not resist heavy rains.
Consequently, the emission reduction due to natural crust formation was low
(0-37%). By adding 4-7½ kg of chopped straw per square metre of tank sur-

face a weather-resistant crust could be formed. This crust could reduce the
NH₃ emission by 63-78%. Compared with the emission reduction due to effective coverings, this was 10% less.

Figure 2 represents the measuring results, and through the individual
results curves have been plotted by linear regression. The curves show the
emission during the storage of cattle slurry in the winter of 1988-1989.
The emission rates from the open tank, the tent and the expanded polystyrene did not change much. The emission from the tanks covered with corrugated sheets and floating foil, however, did increase. With the latter,
this was caused by soiling of the foil. For corrugated sheets it applies
that exchange of air was possible, and the increasing emission rate might
have been caused by an increasing ammonia concentration below the roof.

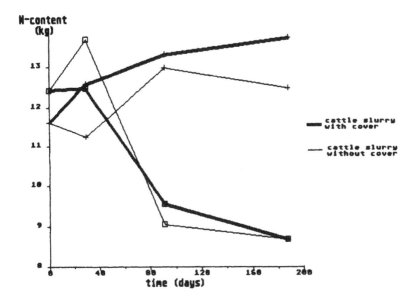

Figure 3: Decomposition of organically bonded nitrogen during storage in
the slurry tank

IN-STORAGE COMPOSITIONAL CHANGES

The nitrogen in the slurry can be distinguished into two main categories: organically bonded nitrogen and ammonium nitrogen. Ammonium is in
equilibrium with ammonia, which is volatile. Volatilization will lead to a
reduction in ammonium content in the slurry, and consequently to a slower
emission rate. It appears from Figure 2, however, that the in-storage
emission rate did not diminish. So, the amount of ammonium in the slurry
must have been replenished, for which decomposition of organically bonded
nitrogen appears to be responsible (Figure 3). During the storage, organically bonded nitrogen was decomposed at a more or less steady rate into
ammonium, volatile fatty acids, carbon dioxide and methane. In an open
tank, the amount of ammonium appeared to remain approximately at the same
level. If the tank was covered, the newly formed ammonium could not vol-

atilize, and this resulted in an increase in ammonium content in the slurry. The manurial value of slurry with a higher ammonium content is more predictable, which makes it more valuable to the farmer.

CONCLUSIONS

- The odour emission from stored slurry was low and could even be reduced by means of covering the storage tanks.
- The NH_3 emission from slurry tanks is strongly related to the season. In summer, emission is two to three times as high as in winter.
- The potential emission rate from stored pig slurry was approximately twice as high as from stored cattle slurry.
- Covering slurry tanks can result in 70-90% reductions. If the exchange of air was not adequately prevented, that is between air above the slurry surface and outside air (as was observed with corrugated sheets) the effects of a cover was considerably lower. The corrugated sheet cover performed a 50% efficiency.
- Crust formation on cattle slurry can reduce the emission by 60-70%. To achieve an effective crust, it had to be stimulated by adding 4-7½ kg/m³ chopped straw to the slurry.
- Covering the tanks resulted in an increased concentration of ammonium in the slurry. With a higher ammonium content in the total nitrogen concentration of the slurry, its manurial value is more predictable.

LITERATURE
(1) VETTER, H, & STEFFENS, G. Wirtschaftseigene Düngung. DLG-Verlag, Frankfurt.
(2) PATNI, N.K. & JUI, P.Y. Changes in nitrogen content of tankstored dairy cattle liquid. Animal Research Centre, Ottawa, Canada.
(3) MUCK, R.E. & STEENHUIS, T.S. (1982). Nitrogen losses from manure storages. Agricultural Wastes 4 p. 41-54.
(4) KAHRS, D. (1980). Abdeckung von Flüssigdungbehältern, RKL.
(5) STEFKENS, G. (1988). Minderung der Geruchsstofffreisetzung durch Abdecken eines Güllelagerbehälters mit einem ceno-Geruchsverschluß. Versuchsbericht der Landwirtschaftskammer Weser-Ems.
(6) KLARENBEEK, J.V., HARREVELD, A.Ph. VAN & JONGEBREUR, A.A. (1985). Geur- en ammoniakemissie van leghennenstallen. IMAG report No. 70.
(7) BODE, M.J.C. DE (1989). Emissie van ammoniak en geur uit mestsilo's en de vermindering van emissie door afdekking, Part 1: varkensmest. IMAG nota No. 464.
(8) PAIN, B.F. & KLARENBEEK, J.V. (1988). Anglo-Dutch experiments on odour and ammonia emissions from landspreading livestock wastes. IMAG Research report No. 88-2.
(9) FRENEY, J.R. & SIMPSON, J.R. (1983). Gaseous loss of nitrogen from plant-soil systems. Development in plant and soil sciences. Vol. 9. Nijhoff/Junk. Den Haag.

EXPERIENCES WITH AUTOMATIC AMMONIA ANALYSES

C.J.M. SCHMIDT-VAN RIEL
Institute of Agricultural Engineering (IMAG)
P.O. Box 43
6700 AA Wageningen, the Netherlands

Summary
        The analysis of air samples is an important aspect of NH_3 emission research. This paper describes the developments and experiences of IMAG.

In the initial period (1986-1987), when a few hundred samples had to be analysed, IMAG used spectrophotometry to measure the ammonium content. Spectrophotometry is an easy and inexpensive analysing method. Air samples from livestock houses, tunnels and pastures were passed through an acid (phosphoric acid) to collect the ammonia. The emissions were high, so that analysing by means of spectrophotometry was simple and satisfactory.

The success of IMAG's NH_3 research programme caused the number of ammonia samples to increase from a few hundreds to an expected 3000 to 6000 samples for the period from mid-March till mid-October in 1988. This large quantity, combined with the fact that lower emissions would also have to be detected requiring more accuracy, resulted in the need for a more advanced analysing method, by which large quantities of samples could be handled. After careful consideration, it was decided to use chromatography to determine ammonium, which method is based on HPLC (High-Performance Liquid Chromatography).

Chromatography amounts to separation based on the distribution of components to be analysed into two phases: a mobile phase serving as a carrier liquid which takes the component through the chromatographic system, and the stationary phase, which is the packing material in the column. As a result of the distribution into two phases, separation will occur, and - provided the retention times for the various components do not overlap - the system gives the results and the components can be detected.

Ammonium is analysed by means of High-Performance Liquid Chromatography. Figure 1 shows the layout of a liquid chromatographic system. It consists of the following components:

- a pump (the solvent delivery system) which causes the eluent or mobile phase to flow through the system,
- an injector device which meters the sample,
- a column in which the separation is brought about, and
- a detector connected to a recorder or integrator to process the results.

The separation of monovalent cations, such as ammonium, is performed by means of the column with sulphonic acid as a functional group ($-SO_3H$). The eluent used is a solution of 30 mM nitric acid.

It is the auto-sampler which can inject 96 samples, which makes the system so effective. The samples can be prepared during the day and the measurements carried out in an automatic process at night. With the help of the integration program Maxima 820 the computer can calculate all the chromatograms the next day, and correct these with regard to the base line etc.

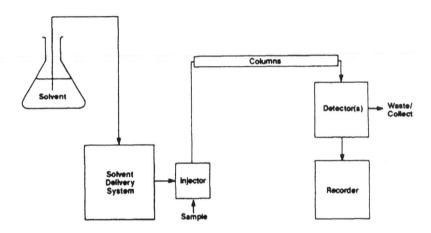

Figure 1: Diagram of the Liquid Chromatographic System

The air samples are passed through a 20 mM nitric acid solution, and the vials are delivered to the laboratory in crates. Here, they are marked with numbers and placed in the sample carriage, after which they are ready to be analysed at night.

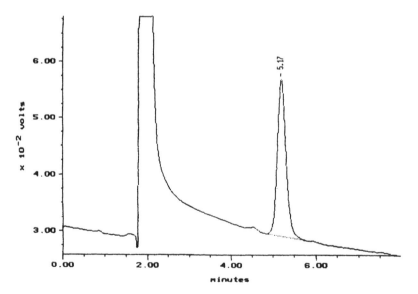

Figure 2: Chromatogram of the standard NH_3, with a peak for NH_3 at a retention time of 5.17 min

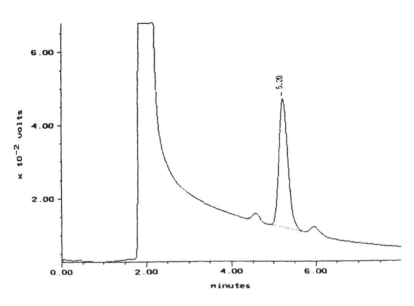

Figure 3: Chromatogram of an air sample, where the ammonia peak can be clearly seen at 5.20 min

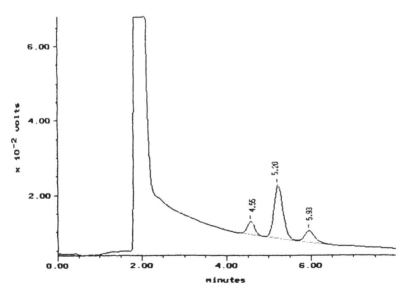

Figure 4: Chromatogram of a sample transported in glass vials, with additional peaks for Na⁺ (at 4.55 min) and K⁺ (at 5.93 min)

The use of this very expensive High-Performance Liquid Chromatography system is not restricted to ammonium analyses. This principle can be used for separating and measuring almost every component. For each type of compound a specific column is available. The only condition is that the component is a liquid or is soluble or adequately emulsifiable in a liquid.

In the past few months slurries and effluents have been analysed for their contents of chloride, nitrate, sulphate, magnesium and calcium. Further, research is being carried out to determine dihydrogen sulphide (H_2S) contents in air.samples.

Modelling the dispersion of ammonia within animal houses

K.-H. Krause[1]
Institute of Biosystems Engineering (FAL)
Braunschweig, FRG
J. Janssen[2]
Institute of Technology (FAL)
Braunschweig, FRG

INTRODUCTION

Several steps concerning the hardware and software for modelling the dispersion of ammonia within animal houses are incomplete. The hardware means the measurement technique of ammonia, the software refers to the numerical prediction of indoor concentration fields. Therefore, co-workers from different divisions of the Institute of Biosystems Engineering[3] from the FAL take part in the solution of this research task.

1. OBJECTIVES

The main objective is the minimization of ammonia emission from animal houses.

The air flow in mechanically ventilated animal houses is highly turbulent. The ventilation rate is determined by physiological data with respect to the animals. Turbulence is induced by ejecting fresh air into the slow-moving air within the animal house and by eddy shedding behind internal obstructions and solid objects. The dispersion of ammonia in this enclosed flow system is determined by the diffusion coefficients or the so-called eddy diffusivities D_x, D_y, D_z. These are the parameters for the turbulent mass flux vectors. The eddy diffusivities result from time-averaging of the product of velocity components $\vec{v} = \vec{v}(u, v, w) = \vec{v}(U + u', V + v', W + w')$ in the direction of an Eulerian coordinate system and concentration $c = C + c'$ of ammonia, where the capitals characterize mean values, while the letters with an apostrophe mark fluctuation terms:

$$\overline{u'c'} = -D_x \frac{\partial c}{\partial x}, \qquad \overline{v'c'} = -D_y \frac{\partial c}{\partial y}, \qquad \overline{w'c'} = -D_z \frac{\partial c}{\partial z}$$

A prerequisite for modelling is the ability to measure the time-dependency of concentration. Only then you may be certain of developing a realistic design methodology for the minimization of ammonia emission based on computational fluid dynamics.

2. FLOW FIELDS IN ANIMAL HOUSES

The ammonia concentration field is described by the species continuity equation. To solve this equation it is necessary to know the air velocity field and the eddy diffusivities for the turbulent transport of ammonia.

Ammonia is transported from the lying area of the animals and from the manure pit through the slotted floor area into the airspace, **Fig. 1.** *The mechanism of this transport is determined by the velocity and the direction of the air mass movement within the livestock building. The velocity vectors depend on the boundary conditions for inlet and outlet cells.*

1 Dr.-Ing. Karl-Heinz Krause belongs to the scientific staff of the Institute of Biosystems Engineering (Director: Prof. Dr.-Ing. A. Munack) and

2 Dr.-Ing. Jan Janssen to the scientific staff of the Institute of Technology (Director: Prof. Dr.-Ing. W. Baader) within the Federal Agricultural Research Centre(FAL).

3 We are very much obliged to Mister Hake, Mack, Pardylla, Speckmann and Zielstorff for their support.

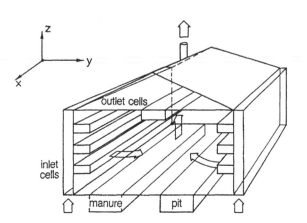

Fig.1 Ventilation system for livestock structures.

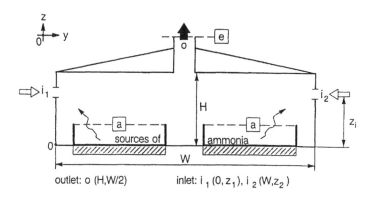

outlet: o (H,W/2)   inlet: $i_1 (0, z_1), i_2 (W, z_2)$

Fig. 2 Cross section of an animal house for the simulation of ammonia transport by suction ventilating.

We regard a system with inlet cells and outlet cells in line. So the recirculating flows in the cross sections of the room are two-dimensional, **Fig. 2**. This consideration reduces the consumption of time for the numerical simulation of fluid flow under different boundary conditions. Furthermore it is possible to verify the calculated results by simple model experiments.

The model for flow visualization is a flat rectangular chamber of perspex. The cover is transparent and can be taken off. The length of the chamber is 25 cm, the height is 5 cm and the width is 0.5 cm. The ceiling of this chamber and the sides are surrounded by channels. In these channels moveable pipes with a hole are installed.

By the moveable pipes it is possible to localize different positions of inlet cells. The outlet cell, that means the suction cell of the ventilating is in the middle of the ceiling. At the bottom four different rectangular flow obstacles can be arranged by magnets.

The flow medium is a mixture of water and glycerin with dispersing aluminium powder. The dynamical similarity between original air flow in the cross section of a livestock building and in the water model requires a velocity of water which is a hundred times higher than the velocity of air. However the geometrical similarity of the produced flow pattern justifies such model experiments at lower Reynolds numbers.

Fig.3 Flow visualization.

When the fluid flows into the chamber from inlets in the ceiling, four characteristic eddies are formed. The visualization, **Fig. 3**, is based on the suspension method. Before movement is initiated solid aluminium tracer particles are distributed uniformly in the fluid. The paths of the tracer reveal the state of the flow.

A pathline is characterized by the curve which shows where a given fluid element has been at earlier times. A stream line is defined as the tangent curve to the velocity of different fluid elements at a certain time. Streamlines and path lines coincide if the direction of the velocity is independent of time.

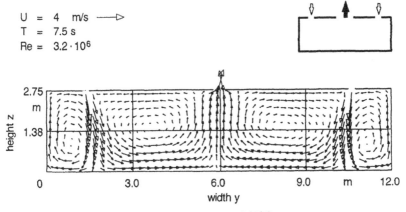

Fig.4 Velocity vector field [1].

The direction of the velocity can be seen from the vector field in **Fig. 4**. The length of a vector stands for the total of velocity, the arrow indicates the flow direction. The velocity field is the solution of the system of differential equations for the balance of mass and momentum. This system is solved by a finite difference technique, called Marker-and-Cell (MAC) computing method. The flow region is covered with a rectangular mesh of cells.

The velocity conditions are set at the outlets only. This is important to be mentioned: livestock buildings are dominated by suction ventilation. That means that we can set the boundary conditions in the outlet cells but the velocities in the inlet cells must be a result of the dynamic flow system.

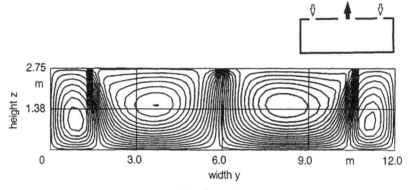

Fig.5 Streamlines.

The lines everywhere tangential to the velocity vectors form a field of streamlines, **Fig. 5**. The space between two streamlines is inversely proportional to the local velocity. A convective transport of mass happens in the stream tubes. A transport perpendicular to the streamlines may occur by diffusion.

Fig.6 An example of flow visualization within a cross section with obstacles.

Fig. 6 shows a photograph of pathlines, **Fig. 7** presents the calculated streamlines. This comparison demonstrates the effectiveness of the computational fluid mechanics. The inlet cells are located in the side walls.

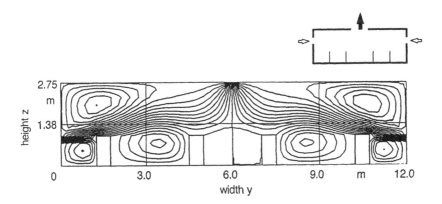

Fig.7 Streamlines.

3. MEASUREMENT OF AMMONIA

Till to-day the visualization of ammonia concentration fields is possible only by evaluation of measurement data. Ammonia sensors allow the registration of time series at different positions in livestock buildings. This is a relative new kind of measurement of ammonia: we measure ammonia concentration in animal houses by means of field effect transistors.

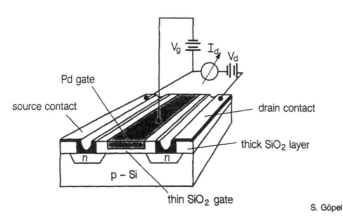

S. Göpel

Fig.8 Construction of an adsorption MOS-field effect transistor.

*The sensor is an advanced development on the basis of the hydrogen sensitive Palladium-MOS capacitor. The sensor utilizes electric effects arising from the catalytic adsorption of gas molecules on the surface of thin active metal gates (Palladium) on top of a semiconductor, **Fig. 8**. The voltage V_g is a measure for the concentration of ammonia.*

The sensor of Sensistor AB has a diameter of 5 mm. The detection range of the sensor is 1-1000 ppm with a response time smaller than 1 s. The accuracy in the range between 4 and 500 ppm is ± 10%.

Primary produced to discover leakages, a careful calibration of the sensors is necessary before application to animal houses.

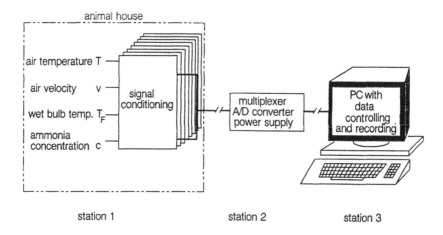

Fig.9 Scheme of the measurement equipment.

*The measurement equipment is divided into three parts, **Fig. 9**. In station 1 the actual signal reception and conditioning happens. Air temperatures and air velocities, wet bulb temperature and ammonia concentrations are measured. The cable length of about 15 m between measuring head and signal conditioning allows the positioning of the measuring head in or near the animal boxes. The signal conditioning unit can be installed in a less dangerous region. Station 2 is connected to station 1 by a 35 m long cable and contains the multiplexer, A/D converter and power supply. The measuring unit is controlled by a personal computer which is also used as data collector in station 3. Due to the distribution, the measuring equipment is very flexible and can be employed even in dusty atmospheres, as sensitive tools like the PC can be installed outside the animal house.*

In practice five sensors were located in an animal house for 37 days [2]. The sensors measured the ammonia concentration in different heights above the floor during 24 hours the day. The vertical air velocity was registered in the outlet cell. At that time our aim was to test the practicability of such a technique. So we limited the expenditure. Meanwhile we have increased the number of our sensor elements.

*The variations of ammonia concentration with time are shown at different height, **Fig. 10**. The uppermost curve stands for the concentrations near the floor, the undermost curve shows the concentration in the outlet. The time series for five different heights are plotted.*

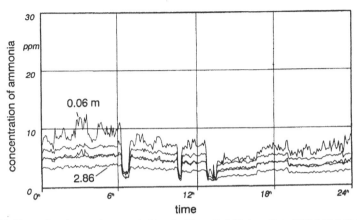

Fig. 10 Time series of ammonia concentrations in different height above floor (2.86, 2.16, 1.46, 0.76, 0.06 m) on 08-01-1989.

Another manner of representation is given in **Fig. 11**. The variations of ammonia concentration with height are shown. Each line belongs to one point of time. The time step is 5 minutes. We observe different concentration gradients from the bottom to the ceiling. The greatest gradients are caused by great ventilation rates. During the night the ventilation is low. The curves in the left part have a smaller gradient in the floor zone than the curves in the right part.

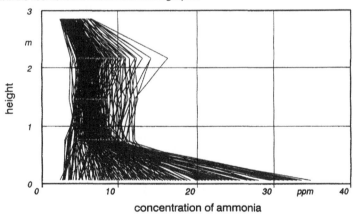

Fig. 11 Variations of ammonia concentrations with height every 5 minutes on 08-16-1989.

78

4. MODELLING THE DISPERSION

"Modelling the dispersion of ammonia" is the trial of linking the air velocity field and the concentration field of ammonia. In atmospheric turbulence the diffusivity in wind direction can be neglected because of the dominance of the convective transport.

No flow direction is favoured in animal houses. Setting the diffusivity to zero, ammonia is transported to the airspace by convective flow. At the bottom we have ammonia sources with the source concentration of 100 ppm. The transport into the air is proportional to the velocity. The isoline for the concentration 10 ppm runs from the left to the right side in the height of 1 m up and down. To demonstrate the influence of the eddy diffusivity the range between 1 and 10 ppm is hatched. In **Fig. 12** the eddy diffusivity is 0, in **Fig. 13** the eddy diffusivity is 1 m^2/s.

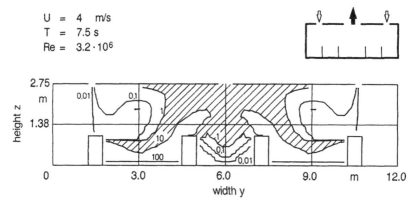

Fig.12 Concentration field with $D_y = D_z = 0\ m^2/s$.

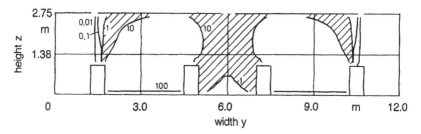

Fig.13 Concentration field with $D_y = D_z = 1\ m^2/s$.

5. CONSEQUENCES TO EMISSION

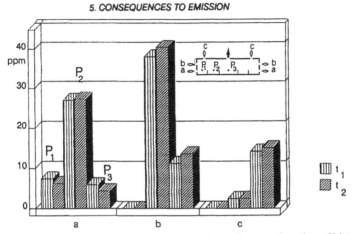

Fig. 14 Ammonia concentration at different positions (P1, P2, P3) for different configurations of inlet cells (a-c).

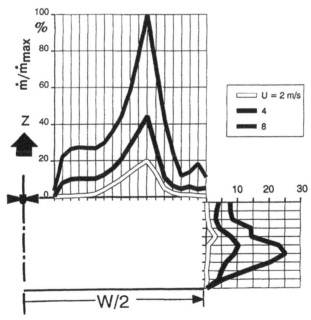

Fig. 15 Relative emitted mass stream in dependence of the positions of the inlet cells.

We suppose that eddy diffusivities in y- and z-direction with 0.01 m^2/s are acceptable. On these premises we simulated different flow situations. The positions of inlet cells vary in the side walls and in the ceiling. We want to know the mass of ammonia emitted at the outlet cells into the environment. Further in connection with this question we want to know whether it is possible to conclude from indoor measurements to the output of ammonia.

We sample the concentration of ammonia at the points P_1, P_2 and P_3 for three configurations of inlet cells, a, b, c at random instants of time t_1 and t_2. This scheme is skeletonized in the pictogram of **Fig. 14.** The differences caused by time can be neglected in comparison with the local differences. The local differences indicate no tendency for estimation of possible outputs of ammonia.

The situation at the outlet cell is studied extensively under varying inlet conditions and suction velocities. The results elucidate that there are maxima of ammonia emission, cf. **Fig. 15.** The inlet cells in the ceiling cause a maximum output at the outer part. The inlet cells in the side walls give maximum output for cell positions above the obstacles. It is worth while to be mentioned that these maxima are found for different suction velocities.

CONCLUSION

There is no homogenous distribution of ammonia within animal houses. The fluid flow pattern must be calculated for different geometrical and suction conditions. To make use of the flow fields it is necessary to condense the results into simplified zone models: compartimentalization models may be a support to realize a control technique. The output of ammonia from animal houses into the environment can be minimized by control of the suction velocity by means of inlet cells.

References

[1] Janssen, J., and Krause, K.-H.: Numerical simulation of airflow in mechanically ventilated animal houses. Building Systems: Room Air and Air Contaminant Distribution, ASHRAE, Atlanta, 1989, 131-135.

[2] Krause, K.-H., and Janssen, J.: Kontinuierliche Ammoniakmessungen in Ställen. Grundlagen der Landtechnik, 39 (1989) 2, 52-65.

Session III

CONTROLLING AMMONIA AND ODOUR EMISSIONS FROM BUILDINGS AND STORES

Chairperson : M. PADUCH

Practical application of bioscrubbing technique to reduce odour and ammonia

S. Schirz, Kuratorium für Technik und Bauwesen in der Land-wirtschaft (KTBL), Darmstadt, F.R.G

Summary

In the past, the problems encountered during operation of bioscrubbers in livestock farms were similar to those with biofilters. Simple and low-priced elements as well as minimized dimensions led to high cost of maintenance and repair. Further developments concentrated on improvements in design details and automation of several operation controls. A two-stage plant was developed especially for ammonia absorption with the aim of simplifying control of the nitrification process. This contribution will present the new elements and discuss the pros and cons.

Introduction

The development of bioscrubbing techniques in livestock farming was exclusively based on odour reduction. It is true, that ammonia in the waste air was always one of the components to be considered in the determination of scrubber efficiency. Recent investigations made by OLDENBURG (1) in a large number of animal houses, however, confirm, the well-known fact (see Figure 1) that high odour concentrations in the waste air do not mean that the ammonia concentrations are high as well and vice versa.

The same is, therefore, true for the efficiency of the bioscrubber. As most of the odorants in the farm waste air dissolve in the scrubbing water as easily as ammonia, the efficiency of the scrubbing system depends on the biological degradation in the packed bed. However, as every group of substances entrains the growth of other strains of micro-organisms, a high reduction in odour does not mean that there is an optimum nitrification of the ammonium forming in the scrubbing water.

Apart from or instead of olfactometric measurements, the determination of the ammonia content in the waste air upstream and downstream of the bioscrubber is often chosen as a method to characterize the efficiency of a bioscrubber in agriculture. This will lead to erroneous results when ammonia is taken as the guide component of odour impressions. The earthy and musty odour downstream of a scrubber may contain ammonia which, however, will not be perceived at a large distance.

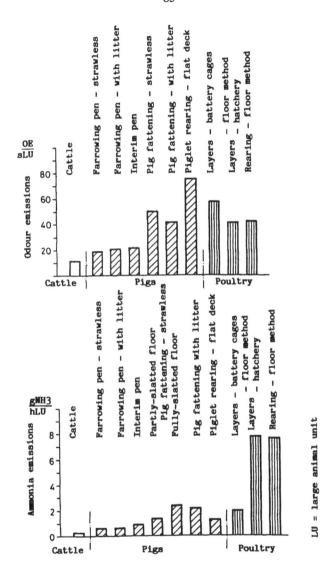

Fig. 1: Comparison of odour and ammonia concentrations of
various species of animals and breeding methods,
according to OLDENBURG (1)

When the efficiency of ammonia reduction of a scrubber is low, this may be attributed to two causes:

- development of nitrification ist not optimum in the scrubbing water, or the absorbing capacity of the scrubbing water is exhausted and ammonia is stripped, respectively,

- the design of the absorber is inadequate or the contacts between water and air need to be improved.

Problems with nitrification were reported by SCHOLTENS et al (2) and DEMERS (3). Their experience provides a good explanation for the continuing variations in pH-values in the system. Moreover, it can be confirmed that also the efficiency of odour reduction achieved with a bioscrubber is at its maximum with a pH-value which can be maintained between 7.0 and 7.5.

Technical problems and further developments of the bio-scrubber as an absorber will be discussed in the following.

Central scrubber for new animal houses

Since 1972, the development of bioscrubbers for application in livestock farms has been simultaneously promoted by IMAG in Wageningen and myself, see SCHIRZ (4). At that time, it was important to design process equipment, which might be installed into existing animal houses and which should also have a very low price. The result were so-called standard scrubbers which could be hung beneath the ceiling of the animal house and the function of which was simple, see SCHIRZ (5). Due to the packaged design of low dimensions, the shape of the packages had to be highly sophisticated. However, they tended to get plugged with sludge and thus caused high cost of maintenance. As a scrubber was designed for approx. 8,000 m³/h (waste air), there were several of them in one animal house - one scrubber each for 100 fattening pigs. In addition to that, the circulation pumps for scrubbing water, the water inlet and the sludge tap turned out to be weak elements. In order to achieve improvements in these respects without increasing the cost involved with bioscrubbing, a central scrubber was developed for new pigpens. It has the following characteristics (Figure 2):

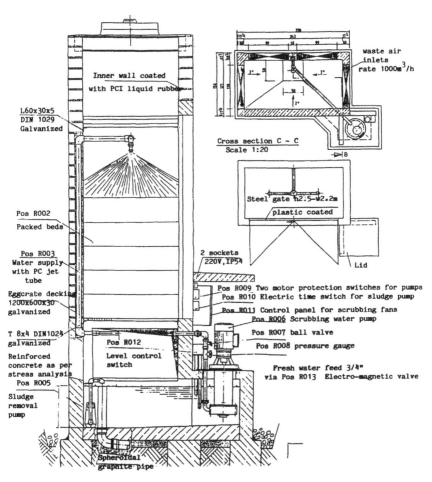

Fig. 2: Central scrubber for waste air flow rates between
20,000 m³/h and 40,000 m³/h

1. Scrubber house
   Masonry of sand lime bricks (17.5 cm), internal
   waterproof plaster, waterproof inspection door.

2. Volume flow rate
   The scrubber is designed for summer flow rates of
   between 20,000 and 40,000 m^3/h. The scrubber house,
   which has internal dimensions of 1.2 m x 2.4 m, can be
   adapted to the different flow rated by means of the
   quantity of installed packings. The pen air is conveyed
   beneath the floor, which means that, in new pens, fans
   can be installed at the end of the waste air duct
   upstream of the scrubber. Additional fans are required -
   see Figure 2 - when a scrubber is installed later on.

3. Packings
   The packings consist of blocks of folded PVC sheets,
   dimensions 0.3m x 0.3m x 1.2m, which are characterized
   by a particularly low pressure loss;

   - effective surface for absorption 110 m^2/m^3
   - construction height 1.8 m (= 6 layers) to 3.6m
     (= 12 layers)
   - water flow rate 5 m^3 (m^2 h)
   - pressure loss 100 to 200 Pa

   The packings are to be installed in dislocated order
   (see Figure 3).

4. Pump for scrubbing water
   The specific design of a submersible pump is used
   like those in scrubbers in the chemical industry
   (see Figure 4);

   - energy consumption 1.1 kW
   - quantity of recirculated water 15 m^3/h

   The pump is resisting corrosion because it is made of
   massive plastic material without metal bearings. The
   circulating quantity of scrubbing water is 2.5 m^3.

5. Nozzles for scrubbing water and droplet separator
   Four maxi-pass nozzles (Bete, USA) operate without
   plugging and spray the packed bed homogeneously;
   - type MP 500, spraying angle 120°.
   A droplet separator is installed above the nozzle
   assembly. It is made of PVC (Munters, Sweden).

6. Sludge pump and water control
   The settling activated sludge is removed with a submersible stainless-steel pump (Jung, W-Germany) and
   pumped into the slurry tank. The quantity of sludge
   to be pumped per day is preset with a time-switch;
   - delivery rate 5 m^3/h
   - quantity of removed sludge 0.2 m^3 to 0.5 m^3 per day.

Fig. 3: Packings 0.3 m x 0.3 m x 1.2 m, made of PVC,
        dislocated installation

Fig. 4: Corrosion-resistant scrubber pump of massive
        plastics, no metal bearings

When the water surface sinks below a defined marker due
to evaporation or sludge removal, a level control switch
will initiate the addition of fresh water by means of a
magnetic valve. The fresh water inlet is cut off when
a second marker has been reached. Both of the control
levels are linked with an alarm system.

The packings turned out to need particularly little
maintenance, i.e. they do not plug, and maintenance
of the plant is required only once a year.

The pH-value must be checked regularly (every third
day), and it should be 7.0 to 7.5. Checking is done most
easily with paper strips (e.g. Lyphan 6.6 zo 8.0).
Automatic pH-measurements (with registration, if
required) is applicable only if the probes are
standardized at regular intervals.

So far, the pH-value in the scrubbing water has been
stabilized by means of sludge removal and fresh water
supply. Following the advice of IMAG, Wageningen,
stabilization is now tested with the addition of lime.
Lime is supplied in the form of granules and placed
in stainless-steel baskets on the packed bed into the
flow from a scrubber nozzle. The granules disslove
slowly and have to be refilled at two or three weeks'
intervals. The consumption amounts to approx. 10 to
30 kg per month. The method is not yet mature because
the granules dissolve in differing velocity depending
on their grain size. The results seem to be positive
as the efficiency of odour reduction could be maintained
at 90 %.

Subsequent installation of a two-stage scrubber

There have been and there will be animal houses into
which scrubbers have to be installed later and in which
an air collection into a central scrubber is impossible
or uneconomic. It is recommended in these cases to
cover each fan with an absorber (Figure 5) and to re-
generate the scrubbing water in a separate bioscrubber
(Figure 6). The absorber is designed to match with the
air flow of the pertinent fan and operates as a spray
scrubber with packed bed, Figure 7. Two rings of 8
maxi-pass nozzles each, type MP 250, spray in opposite
directions. In the section between the air cone above
the fan and the nozzles, there is a rectifying device
made of perforated tubes, dia. 40 mm. So far, problems
occurred with the nozzles which were plugged by too
many dead flies in the scrubbing water. A fine screen
upstream of the outlet socket improved the situation.

Fig. 5: Separate absorber on each fan of the animal house

Fig. 6: Central bioscrubber into which the water is
conveyed from several absorbers

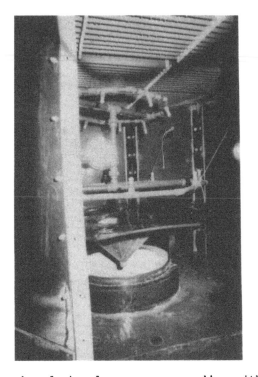

Fig. 7: Absorber designed as a spray scrubber with packings

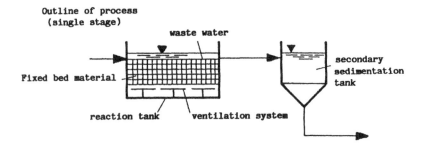

Fig. 8: Scheme of submersed packings, according to ENVICON

The central scrubber is designed for between 8 and 10
absorbers with a waste air capacity of approx. 5,000 m³/h.
The scrubbing water ist pumped into all of the absorbers
via a pump with a power of 3.1 kW and is then conveyed into
a central tank. Dust and sludge are precipitated in several
chambers. Afterwards, the water is sprayed across the
biolayer of the packages by means of a separate pump. The
constructional expenditure of this scrubber is rather high
so that considerations were made to use a submersed packing
instead, see Figure 8. This scrubber, however, would require
pressure ventilation, while the other one can operate with
a fan for an air flow rate of 500 m³/h.

The efficiency of odour reduction depends on the absorbers
and comes up to approx. 80 %, which is less than in the
central scrubbers.

This plant was designed to utilize the advantages of a
central plant also for subsequent installations in existing
animal houses.

As absorbers and central scrubber operate with separate
water circulation, the water may be conditioned. The next
test is aimed at acidifying the scrubbing water in order
to increase the cleaning efficiency. When the second stage
is then maintained in an alkaline state with the addition
of lime granules, this might well turn out to be advantageous
for nitrification in the scrubbing water.

Literatur

(1) Oldenburg, J.: Ammoniakemissionen und -immissionen aus
    der Tierhaltung, KTBL-Schrift 333, Darmstadt 1989

(2) Scholtens, R et al.: Control of amonia emissions with
    biofilters and bioscrubbers. In: Voltaile emissions from
    livestock farming and sewage operations, S. 196-208.
    Elsevier London /Ny 1988

(3) Demmers, T.G.M.: Adsorption und Nitrifikation von
    Ammoniak im Biowäscher, In: Biologische Abgasreinigung
    VDI-Bericht 735, S. 147-160, Düsseldorf 1989

(4) Schirz, S.: Abluftreinigungsverfahren in der Intensiv-
    tierhaltung, KTBL-Schrift 200, Darmstadt 1975

(5) Schirz, S.: Design and experience obtained with
    bioscrubbers, In: Odour prevention and control of
    organic sludge and livestock farming, S. 241-250
    Elsevier London/Ny 1986

BIOFILTERS AND AIR SCRUBBERS IN THE NETHERLANDS

R. SCHOLTENS & T.G.M. DEMMERS

Institute of Agricultural Engineering (IMAG)
P.O. Box 43, 6700 AA Wageningen, the Netherlands

Summary
    For more than a decade the technique of treating exhaust air from livestock houses through biofilters has been known. Nonetheless, biofilters are hardly used in intensive livestock farming, for which there are a few reasons. One is the cost factor of exhaust air treatment, in particular with a view to the large air quantities involved. Another cost factor is the handling of effluent water needed to remove the nitrogen absorbed in the biofilter material. A third factor is the performance in the daily practice which does not always come up to expectations.
    Initially, air scrubbers were used for odour control, mainly on farms which were forced to do so under their Nuisance Act licence. In the last few years, acid rain has become a problem, and this has caused the control of NH_3 emission to become a key item. Air scrubbers are now used to control both ammonia and odour emissions. The disposal of effluent water from bioscrubbers is a problem in the actual application.

1. PRINCIPLE OF AN AIR SCRUBBER

    An air scrubber is a packed-bed unit through which exhaust air passes. Water trickles and circulates through the packed bed, which has been designed to ensure the best possible contact between water and air. In the water and on the packed bed a biologically activated sludge develops. The ammonia absorbed in the process water is first converted into nitrite and then into nitrate, which is called nitrification. The odour components absorbed in the same process are decomposed into CO_2 and H_2O. As a result the process water can again absorb ammonia and odour species.
    To prevent the activity in the activated sludge to be curbed by accumulating nitrogen concentrations, the process water has to be replaced at regular intervals. In modern air scrubbers, the nitrogen content is controlled by draining a certain amount of the process water as effluent.
    Ammonia continues to be absorbed at a pH value lower than 7.8 and when the ammonium concentration is low. Nitrification, however, is a sensitive microbiological process.

2. NITRIFICATION

    Nitrification is the oxidation of ammonia to nitrate. It takes place in two stages, first into nitrite, then into nitrate. Two different groups of autographic bacteria are responsible for the conversions, namely Nitrosomonas spp. and Nitrobacter spp.

$$NH_4^+ + 1\tfrac{1}{2} O_2 \rightarrow NO_2- + 2H^+ + H_2O \qquad \text{(Nitrosomonas)}$$
$$NO_2- + \tfrac{1}{2} O_2 \rightarrow NO_3- \qquad \text{(Nitrobacter)}$$

    The optimum pH value for nitrifying bacteria is between 7 and 8. Below pH 6.0 and above pH 10.2 the nitrification process stops altogether. When Nitrosomonas forms nitrite, acid is also produced. This acid has to be neutralized in order to maintain the optimum pH range. If there are no neutralizing substances (alkalinity) in the process water, they have to be added. These can be pH–controlled hydrogen carbonate, a caustic solution or solid substances with buffer capacity.

Adding marble has proven to facilitate the nitrification process well at a value of pH 4.5. This is possible because of the biofilm which develops on the marble surface. The pH value inside the layer close to the marble is considerably higher than the value in the effluent water.

Research has shown that it is not ammonium and nitrite but the free ammonia (fa) and the free nitrous acid (fna) which inhibit the nitrification process. The free ammonia inhibits the Nitrosomonas bacteria in concentrations of 10–150 mg/l, whereas the Nitrobacter bacteria are inhibited in concentrations of 0.1–1.0 mg/l. Free nitrous acid inhibits the two organisms in concentrations of 0.22–2.8 mg/l. Free ammonia and free nitrous acid concentrations highly depend on the pH value. If it rises, the concentration of free ammonia increases and that of free nitrous acid diminishes. In case of a high pH value, free ammonia inhibits the nitrification process, and if it is low, free nitrous acid does so.

In addition to the free ammonia and free nitrous acid, the nitrate produced can inhibit the nitrification process. If activated sludge is suddenly exposed to nitrate, nitrification is severely inhibited. In a concentration of 0.5 g NO_3^-– N/l, nitrification is inhibited by 50%. If activated sludge is slowly adapted to high nitrate concentrations, nitrification is not even inhibited in a concentration of 2 g NO_3^-–N/l. A concentration of 2.7 g NO_3^-–N/l will cause a 50% inhibition, and if the concentration is between 4 and 5 g NO_3^-–N/l, nitrification stops altogether.

Accumulation of nitrite in activated sludge (incomplete nitrification from nitrite to nitrate) can occur under the following circumstances:

– lowered temperature
– limited presence of O_2 and/or CO_2
– increased pH value
– presence of free ammonia
– increased removal or loss of sludge
– acute high burden of ammonia

The formation of nitrite entails considerable risk, as nitrite is toxic to humans and animals. To humans nitrite is much more toxic than nitrate. The fatal dose of nitrite is 3 mg/kg body weight, whereas the same of nitrate is 3g/kg body weight.

3. AIR SCRUBBERS IN PRACTICE

The performance of cross–flow air scrubbers have been measured. It has become clear that the present air scrubbers do not adequately remove ammonia. The efficiency appears to be closely related to the amount of effluent water. If the effluent flow is below the 'critical' value, the efficiency diminishes. This is due to the fact that a certain quantity of effluent water can remove only a limited amount of nitrogen. The air scrubbers efficiency depends on the removal of absorbed nitrogen.

As a result of the low alkalinity of tap water, the pH value of the process water falls as soon as nitrification starts. The low pH value facilitates a higher absorption of ammonia, whereas the rate of nitrification remains the same. This entails an accumulation of ammonia, and the presence of free ammonia causes the nitrification process to slow down. As a result, nitrification stops at the stage of nitrite, with the rate being reduced because of the low pH value.

The decrease in pH value is compensated only by the absorption of ammonia. To compensate the formation of 1 mg NO_2-N, an amount of 2 mg NH_4-N has to be absorbed. this means that the process water always contains an amount of ammonia which is equivalent to nitrite.

The process water contains ammonia and nitrite concentrations up to 2 g N/l. These components make the effluent very toxic. This effluent can only be drained into the sewerage system or to the slurry pit. The large amount does not make disposal to the slurry pit an interesting option, but in many cases there is no other solution.

4. TESTS WITH SEMI-TECHNICAL AIR SCRUBBERS

In order to remove all ammonia from the process water, the pH value has to be maintained at an optimal level for nitrification, so that no unduly low pH value and free ammonia will cause the process to slow down.

Four semi-technical air scrubbers have been used in the research into the complete removal of ammonia from effluent water. The aim was to investigate how the nitrification process can be optimized. It has proven to be ineffective to add hydrogen carbonate to stabilize the pH value at 7.0, which is appropriate for ammonia absorption. Despite several interventions, and an increase in effluent water, the nitrification process would not start. Not until the pH value was raised up to 7.3 was there a proper nitrification process producing nitrate. The nitrification rate could be raised up to 25 mg N/m^2 h. The load then was 0.6 g N/h, the efficiency being 95–99%. With a value of 0.13 l/h, the amount of effluent water was low.

The nitrification rate at increasing load did not increase because of the high nitrate concentration. As a result the ammonium concentration and the amount of free ammonia accumulated. This prolonged overload interfered with the nitrification process, which after some time did not go beyond the stage of nitrite.

In a next experiment the pH value was stabilized by adding marble to the packed bed. This resulted in a proper start of the nitrification process. The pH value varied between 6.8 and 7.5. The ammonia and nitrite concentrations were low. When the ammonia load was increased, however, the alkalinity of the marble appeared to be insufficient. The pH value fell, and because of this and the increasing concentrations of ammonium and free ammonia the nitrification rate was only 12 mg N/m^2h. After the load had been reduced, the nitrification process appeared to be disturbed. At a low pH value of 5.8 much nitrite was formed in addition to nitrate. The ammonium concentration remained low at the lowered load.

The fact that there was continuous formation of nitrite in addition to nitrate (contrary to the other air scrubber), is attributed to the nitrifying bacteria living in the biofilm on the marble. The buffering capacity of the marble protects them against the effects of the pH value in the process water. The biofilm, however, has an unfavourable effect on the buffer capacity of the process water, as it interferes with the exchange of substances between process water and marble. The surface of the marble is too small for the nitrification process to take place in the biofilm on the marble only, so that buffering of the process water has to be performed in a different way.

By raising the pH value up to 7.3, by means of a pH controller, the nitrification process could be fully restored and the load be increased again.

It is possible to improve the biological processes in the air scrubber. In particular, the control over the pH value is a considerable improvement. It has not been decided, which method will be selected.

The air scrubber will have to be adapted. The maximum nitrification rate found per unit area does not suffice to convert all absorbed ammonia. This requires the surface of the packing to be enlarged by 15 times. It has to be remarked here, that the maximum nitrification rates were measured in the semi-technical air scrubbers, whereas the nitrification process was inhibited by nitrate.

With a view to their toxicity, the concentrations of free ammonia and free nitrous acid must be kept low. This can be achieved by considerably increasing the volume of process water in the air scrubber so that sudden increases in concentrations are levelled.

The effluent water remains one of the major disadvantages of the air scrubber. The amount of effluent can be reduced in various ways, e.g. by means of de-nitrification and by reverse osmosis to upgrade the effluent water. Both processes offer good prospects and will be subjected to further research.

5. MEASUREMENTS OF BIOFILTERS IN PRACTICE

TNO has performed practical research to investigate the effects of a biofilter installed in a pig feeding house. The biofilter had been designed by IMAG. It had two separate compartments. One compartment contained compost/bark as a biological filter, whereas the second contained peat/heather for the same purpose. The dimensions of the two compartments were based on the results of the two previous stages of this research project.

The major results of the experiment are as follows:

- Peat/heather and compost/bark can be used as biological filters for pig feeding houses. Of NH_3 at least 85% was removed, and the reduction in odour emission was at least 75%.
- For a proper functioning the filters should be moistened, well at regular intervals.
- In order to reduce the pressure head across the filter, it should preferably be turned over at least once every four months.

Filter material	Unit	Peat/heather	Compost/bark
Maximum surface load	$m^3/m^2.h$	300	450
Height of bed	m	0.5	0.5
Maximum pressure head across filter medium	Pa	≤ 120	≤ 250
Maximum pressure head across dust filter	Pa	≤ 40	≤ 40
Required surface of filter medium per pig	m^2	0.33	0.23
NH_3 removal	%	≥ 85	≥ 85
Odour removal	%	≥ 75	≥ 75

Table 1: Guidelines for biofilters

On the basis of the practical research carried out at the Research Station for Pig Farming (PV) in Rosmalen guidelines can be drafted for the dimensioning of biofilters. For dimensioning and drafting biofilters for pig feeding units, the guidelines can be applied as given in the table.

On the basis of the results of the experiments, the technical and financial assessment of design, construction and management of a biofilter in intensive farming have been made. IMAG has performed this assessment to conclude the third stage of the research project.

It was found that the application of biofilters will cost Hfl 12–19 per pig delivered, depending on the suitability of the piggery to be equipped with a biofilter. The costs of the two filter packed bed are more or less identical.

To the costs stated, the costs of draining of the effluent and cleaning water have to be added, which so far can only be estimated, as it is not known how much process water will have to be drained. The estimate varies from under Hfl 4 to more than Hfl 8 (based on drainage to the slurry storage, depending on housing type). So the total estimate for biofilters in pig feeding houses ranges from under Hfl 16 to more than Hfl 27.

REFERENCES
(1)   DEMMERS, T.G.M. & SCHOLTENS, R., 1988. Bestrijding van ammoniakemissies met biowassers, IMAG Nota No, 422, 43 p.
(2)   EGGELS, P.G. & SCHOLTENS, R., 1989. Biofiltratie van NH_3–bevattende stallucht bij de intensieve veehouder, Fase 3: Onderzoek aan een biofilter op praktijkschaal alsmede consequenties voor biofiltratie in de praktijk, Ref.–No. 89–107, 72p.

Session IV

AMMONIA AND ODOUR EMISSIONS FROM LAND SPREADING MANURES

Chairperson : B.F. PAIN

FACTORS INFLUENCING THE ODOUR AND AMMONIA EMISSIONS DURING AND AFTER THE LAND SPREADING OF ANIMAL SLURRIES

V R PHILLIPS[1], B F PAIN[2], J V KLARENBEEK[3]

1. AFRC Engineering, Wrest Park, Silsoe, Bedford, MK45 4HS, UK
2. AFRC Institute for Grassland and Environmental Research, Hurley, Maidenhead, SL6 5LR, UK
3. IMAG, Postbus 43, 6700 AA Wageningen, Netherlands

SUMMARY

Two different types of experiments were mounted to compare the emissions of odour and ammonia from different types of slurry spreading machine. The first type of experiment, in which a Land-Rover equipped with a sampling frame followed behind a spreader, compared emissions arising during the act of spreading. For ammonia emissions, a low-trajectory spreader, a shallow injector (75 mm) and a deep injector (150 mm) all gave significantly lower emissions than did a conventional spreader, but a light-weight irrigator gave higher emissions. However, for odour, because the standard errors of the measurements were greater than for ammonia, there were hardly significant differences between the machines.

The second type of experiment, which used a micrometeorological technique based on large circular plots, compared emissions _after_ spreading, when the slurry was lying on the ground. Relative to a conventional spreader, a low-trajectory spreader gave some reduction in odour emission, but both shallow (75 mm) and deep (150 mm) injection gave a much greater reduction. Since shallow injection was almost as effective as deep injection but is expected to be cheaper to operate, it is recommended that this technique receive further study.

The emissions per m^3 of slurry spread were many-fold less during spreading than after spreading, with the former being less than 1% of the latter.

For one typical kind of farm, the costs per m^3 of slurry spread, calculated for various different systems of spreading which aimed to reduce emissions, ranged between 1.05 and 2.16 times the costs of conventional spreading.

1. INTRODUCTION

In the UK, the land spreading of slurries and manures causes more complaints about odour than does any other stage of livestock production. Also, evidence is growing that ammonia emissions from agriculture are a major contributor to forest die-back, and ammonia emissions from land spreading of slurries are expected to be one of the major sources of ammonia emissions from agriculture. Studies have therefore been mounted to measure emissions from land spreading and to investigate various ways to reduce them, at minimum cost.

The work has been a joint project of AFRC Engineering, IGER, IMAG Netherlands, Silsoe College, ADAS (Ministry of Agriculture) and the Water Research Centre.

Most of the odour concentration measurements were made using "Olfaktomat" olfactometers built by Project Research Co., Amsterdam, and the remainder were made using "IMAG 5" olfactometers. The measurements of ammonia-in-air concentrations were made by drawing known volumes of the air through bubblers, in which excess acid (phosphoric or sulphuric) absorbed the ammonia quantitatively. The acid was then assayed colorimetrically or by ion chromatography.

2. INFLUENCE OF TYPE OF SPREADING MACHINE ON EMISSIONS

2.1 Emissions during spreading

Experiments were mounted to compare the emissions during the act of spreading pig slurry on grassland using five different spreading machines:
a. Conventional vacuum tanker with splash plate (as a control).
b. Low-trajectory surface spreader.
c. Deep injector (winged tines 150 mm deep).
d. Shallow injector (wingless tines 75 mm deep).
e. Light-weight cable-driven irrigator, equipped with curtains to reduce drift.

A measuring technique developed at IMAG was employed, in which the spreading machine is driven directly upwind and a Land-Rover equipped with an array of sampling points on a large frame (Fig.1) follows behind, at the minimum distance which ensures that no droplets of slurry can enter the samplers. Sampling pumps, air-flow meters etc., are all mounted in the back of the Land-Rover.

Fig.1. Land-Rover with its sampling frame

Triplicated runs in random order were made with each machine. Normal day-to-day variations e.g. in the wind velocity, air temperature, air humidity and slurry temperature were thus able to exert their influence on the standard deviations in the measured rates of emission. The target slurry spreading rate was 43 m^3/ha, the mean % Dry Matter was 2.9%, the mean % total-N was 0.36% and the mean % NH_4^+-N was 0.26%. Wind speed as a function of height, and wind direction, were continuously monitored during the period of the experiments and odour and ammonia background levels were measured during each experiment.

The results are summarised in Table 1, in which emission rates are given as both per unit time and per m^3 of slurry spread. The t-test was used to test whether the between-machine differences in the mean observed total emissions per m^3 of slurry spread were significant. The total ammonia emissions per m^3 spread from the irrigator were significantly greater (5% level) than those from the conventional vacuum tanker: and these in turn were significantly greater (5% level) than those from any of the three other machines. There were no significant differences between the ammonia emissions from these three other machines (the shallow injector, the low-trajectory surface spreader, and the deep injector).

Turning to odour emissions, there was only one significant difference between the odour emissions per m^3 of slurry spread, for any pair of machines: the emissions were significantly greater (but only at the 10% level) from the conventional machine than from the low-trajectory spreader.

The differences in the above findings, for ammonia and odour emission rates, are at least partly explained by differences in the standard error of the measurement techniques in the two cases. As Table 1 implies, the olfactometry-based odour measurements have an inherently higher standard error than do the ammonia rate measurements. However, it is not possible to dispense with direct odour measurement and rely on ammonia as an objective indicator of odour, because the correlations found between the two types of measurement are variable. (Proposals are being drawn up for work to identify and quantify the various separate errors which together make up the relatively large overall standard error on odour measurements.)

The ammonia emission rate from the light-weight irrigator was large despite its shrouding curtains. The high emission rate was probably caused by the reliance of its design, on high speed discharge of slurry through two small nozzles - conditions which gave very small droplets of slurry and thus encouraged loss of ammonia and other volatiles from the slurry by evaporation. The shrouding curtains may have prevented drift of liquid droplets but did not greatly reduce the process of volatilization. In any event, the manufacturer has re-designed the nozzle section of the device since our experiments.

2.2 <u>Emissions after spreading</u>

A micrometeorological balance method reported by Denmead (1) was employed, which is based on the use of large circular plots. An earlier experiment using this method has already been reported (2). For convenience, the results of that earlier experiment are re-stated here, in graphical form (Figs. 2.1 and 2.2). It was concluded then that injection generally gave far lower emission rates of both odour and ammonia than did either a conventional or a low-trajectory surface spreader. For example, immediately after spreading, the injected plot gave an odour emission rate which was 27% of that from the low-trajectory plot, and 15% (linearly corrected for minor differences in spreading rate, m^3/ha) of that from the conventionally-spread plot; while, for ammonia emission rates the corresponding ratios were 3% and 4% respectively. The low-trajectory surface spreader gave, corrected, 30% lower odour emission rate, but 50% higher ammonia emission rate, than did the conventional spreader. With all three methods, rates of emission fell rapidly with time after spreading, so that after 24 hours, no rate was more than 17% of the respective initial rate.

However, a secondary peak of odour, 48 hours after spreading, was noted for the deep injector (see Fig.2.1). A further "three circle" experiment, using the same principles and techniques as the earlier one, was therefore mounted, with one objective being to either verify or disprove this secondary peak. The low-trajectory spreader was replaced in this latter experiment by

Machine type and Experiment Number	Odour emission rate, per machine (thousands of OU)/s		Total odour emissions (thousands of OU)/ (m³ of slurry spread)		Ammonia emission rate per machine mg NH₃/s, corrected for background		Total ammonia emissions g NH₃/(m³ of slurry spread) corrected for background	
	triplicates	mean ± sd	triplicates	mean ± sd	triplicates	mean ± sd	triplicates	mean ± sd
Conventional vacuum tanker with splash plate 1 / 2 / 3	10.47 / 6.21 / 6.94	7.87 ± 2.27	496 / 250 / 301	349 ± 130	35.9 / 28.4 / 51.0	38.4 ± 11.5	1.70 / 1.14 / 2.21	1.68 ± 0.54
Prototype shallow injector 1 / 2 / 3	6.11 / 2.05 / 1.74	3.30 ± 2.43	244 / 87 / 67	133 ± 97	6.6 / 6.2 / 3.1	5.3 ± 1.9	0.26 / 0.26 / 0.12	0.21 ± 0.08
Low-trajectory spreader with 15 trailing hoses 1 / 2 / 3	0.15 / 1.53 / 1.46	1.05 ± 0.78	5 / 44 / 57	35 ± 27	0.2 / 12.0 / 3.3	5.2 ± 6.1	0.01 / 0.34 / 0.13	0.16 ± 0.17
Deep injector 1 / 2 / 3	1.95 / 4.54 / 1.53	2.67 ± 1.63	127 / 297 / 121	182 ± 100	0.0 / 1.4 / 2.3	1.2 ± 1.2	0.00 / 0.09 / 0.18	0.09 ± 0.09
Cable-driven irrigator with shrouding curtains 1 / 2 / 3	7.50 / 61.04 / 23.01	30.52 ± 27.5	1,575 / 12,734 / 5,263	6,520 ± 5,690	83.9 / 149.3 / 81.6	104.9 ± 38	17.62 / 31.15 / 18.66	22.48 ± 7.5

Table 1 : Emission rates of odour and of ammonia during the land spreading of pig slurry by the five different types of machine

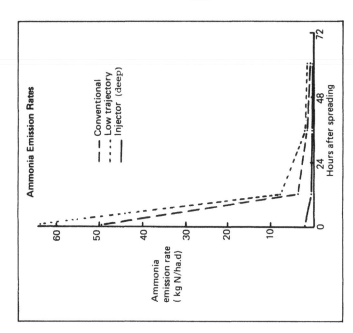

Fig.2.2

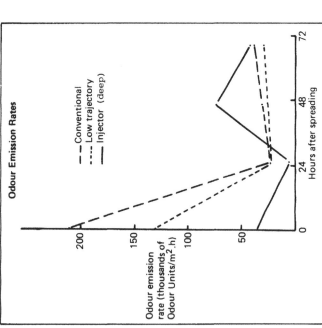

Fig.2.1

Rates of emission as a function of time after simultaneous spreading on three identical plots using three different methods (first experiment)

a prototype shallow (75 mm) injector, a second objective of the experiment being to compare deep and shallow injection.

The area of land, the crop height, the type and rate of application of slurry (100 m³/ha) were all the same as for the earlier experiment.

Table 2 shows the basic odour concentration results. For each of the machines, the odour concentrations drop rapidly within the first hours after spreading has been completed, with the odour concentration immediately after spreading being considerably higher for the conventional spreader than for either the deep or the shallow injector.

	Hours after spreading				
	0	3.5	24.5	45	123
Conventional spreader	59	25	14	17	15
Shallow injector	15	33	no data	8	10
Deep injector	28	7	0	21	19
Background	7	no data	14	28	2

Table 2 : Odour concentration measurements (Numbers of Dilutions to Threshold, or Odour Units/m³ air) as a function of time after simultaneous spreading on three identical plots using three different methods (second experiment)

The odour emission rates were calculated from these odour concentrations via the wind speeds across the plots (micrometeorological mass balance method). The resulting odour emission rates are plotted in Fig.3.1. These odour emission rates, like the odour concentrations, first drop rapidly after spreading has been completed, with the rate immediately after spreading being considerably higher for the conventional spreader than for either the deep or the shallow injector. However, the later behaviour of the emission rates differs from that of the odour concentrations, in that the rates at 123 h after spreading are much higher, again. The immediate reason is that the wind speed was also much greater at 123 h than at all shorter times after spreading (see bottom of Fig.3.1).

The ammonia emission rates, plotted on similar axes, in Fig.3.2, show a similar trend, albeit with a smaller rise at 123 h.

The observation made in the earlier experiment that the odour emission rate showed a secondary peak for injection alone, was not repeated. Considering the emission rates immediately after spreading, the values of these for ammonia from this experiment agreed well with those from the earlier experiment. For odour, there was good agreement in the case of injection, but initial emission rates were considerably lower this time in the case of the conventional spreader. Lower soil temperatures may have contributed to these differences: the later experiment was conducted under much colder conditions than was the earlier one.

The differences between the results from the deep and the shallow injector were never great. The potential of shallow injection e.g. hardly less reduction of emissions than for deep injection, but with considerably reduced draft forces, should therefore be further explored.

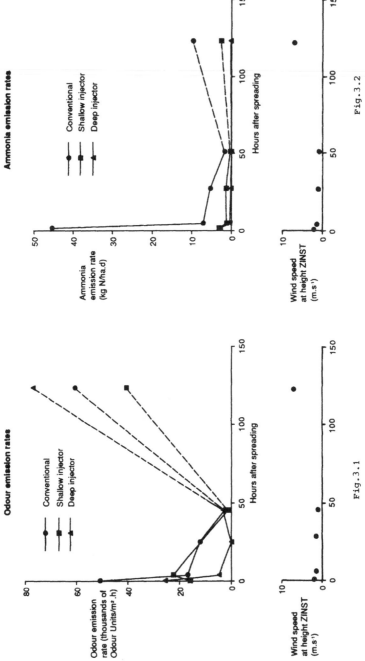

Fig.3.1

Fig.3.2

Rates of emission (and wind speed at height ZINST) as a function of time after spreading on three
identical plots using three different methods (second experiment)

3. THE RELATIVE MAGNITUDE OF TOTAL EMISSIONS, DURING AND AFTER SPREADING

In all cases, the emissions per m^3 of slurry spread were many-fold less during spreading than after spreading, with the former being less than 1% of the latter. Nevertheless, changing from a conventional spreader to a low-trajectory type may well still inherently reduce nuisance during spreading, because of the visual/psychological advantage of the latter type of spreader: with this type, it is far less obvious to neighbours or to passers-by that slurry spreading is under way than it is with the conventional machines which throw slurry droplets high in the air.

4. THE COSTS OF ACHIEVING REDUCTIONS IN EMISSIONS

The costs are, of course, a very important aspect of reducing emissions, but not one which is easy to assess comprehensively. A good start on the assessment of costs has been made by Warner et al.(4), who compared the system costs of spreading on one typical kind of farm, when using a number of different methods which sought in different ways to reduce the emissions from land spreading. However, the calculations of Warner et al.(4) do not make any allowances for the relative efficiencies of the different methods of reducing emissions, nor for such factors as the fertilizer value of the ammonia, the emission of which has been prevented. Warner et al. found that the calculated extra costs (before grant) per m^3 of slurry spread, incurred by altering the method of spreading so as to reduce emissions, ranged between 5% (if a tanker is equipped with a low-trajectory boom but no other changes are made) and 116% (if the slurry is anaerobically digested before spreading).

Work is continuing in the UK and in the Netherlands, to augment and refine this kind of information on the costs of reducing emissions. The ultimate aim is to be able to calculate, for any method, a cost per unit of emission prevented.

AFRC Engineering and ADAS (Ministry of Agriculture) have begun work (5) on a Waste Engineering Expert System (WEES), which will incorporate such cost information, as well as a range of other information on different aspects of the treatment and management of agricultural wastes.

5. CONCLUSIONS

5.1 With the current measuring techniques, measurements of odour concentrations have proportionately greater standard errors than do measurements of ammonia concentrations.

5.2 During the act of spreading, a low-trajectory spreader, a shallow injector and a deep injector all gave significantly lower ammonia emissions, per m^3 of slurry spread, than did a conventional spreader. However, a light-weight irrigator which caused the slurry to be broken up into very small droplets gave higher ammonia emissions than did the conventional spreader.

For odour emissions, the equivalent comparisons were inconclusive, because of the larger standard errors in that case.

5.3 The emissions after spreading were in almost all cases less for low trajectory spreading or for injection, than for conventional spreading, with injection reducing emissions after spreading more than did low trajectory spreading.

Shallow injection (75 mm) gave almost as much reduction as did deep injection (150 mm). Since shallow injection may bring other advantages, such as reduced costs, it is recommended that this technique be studied in more detail.

5.4 In all cases, the emissions per m^3 of slurry spread were many-fold less during spreading than after spreading, with the former being less than 1% of the latter.

5.5 For one typical kind of farm, the costs per m^3 of slurry spread, calculated for various different systems of spreading which aimed to reduce emissions, ranged between 1.05 and 2.16 times the costs of conventional spreading.

REFERENCES
(1) DENMEAD, O.T. (1983) Micrometeorological methods for measuring gaseous losses of nitrogen in the field. In : Freney, J.R. and Simpson, J.R., eds. Gaseous losses of nitrogen from plant-soil systems. pp 133-157. The Hague, Martinus Nijhoff.
(2) PHILLIPS, V.R.; PAIN, B.F.; WARNER, N.L.; CLARKSON, C.R. (1988). Preliminary experiments to compare the odour and ammonia emissions after spreading pig slurry on land using three different methods. In : Cox, S.W.R., ed. Engineering advances for agriculture and food. London, Butterworths.
(3) LOCKYER, D.R. (1984). A system for the measurement in the field of losses of ammonia through volatilization. J. Sci. Food Agric. 35 837-848
(4) WARNER, N.L.; GODWIN, R.J.; HANN, M.J. (1990). An economic analysis of slurry treatment and spreading systems for odour control. The Agricultural Engineer, in press.
(5) BEAULAH, S.A.; BREWER, A.J.; CUMBY, T.R.; HALL, C.A.; PHILLIPS, V.R. (1990). The development of WEES : A Waste Engineering Expert System. Paper presented at the ASAE Sixth International Symposium on Agricultural and Food Processing Wastes, Chicago, 17-18 December.

AMMONIA EMISSIONS AFTER LAND SPREADING OF ANIMAL SLURRIES

J.V. KLARENBEEK & M.A. BRUINS

Institute of Agricultural Engineering,
P.O. Box 43
6700 AA Wageningen, the Netherlands

Summary
As a result of the goal set by the Dutch authorities, the annual ammonia emissions have to be reduced by 50-70% in 2000. Since most of the emissions are animal-based, a great strain is put on livestock husbandry in the Netherlands. Besides livestock houses, slurry storage and grazing cattle, ammonia emissions from land-spread slurries and manures have to be reduced. Possible pathways for the achievement of this are discussed in this paper.

INTRODUCTION

Annually, several thousands cubic metres of slurry are applied as fertilizer in the Netherlands. It is estimated that in 1990 approximately $50 \cdot 10^6$ m³ will be used in land spreading. Besides the positive effects such as higher yields due to the application of cattle, pig and poultry slurries, some negative aspects are known. Pain and Klarenbeek (1988) have established that 1 m³ of pig slurry, when applied, emits $26 \cdot 10^5$ odour units and 2 kg of ammonia.

The total ammonia emission in the Netherlands for 1986 has been estimated at $220 \cdot 10^6$ kg, 90% of which is associated with intensive livestock keeping. Futhermore, it is estimated that just over 50% ($114 \cdot 10^6$ kg) is released during and following land spreading of slurries. To overcome problems of soil acidification, in which atmospheric ammonia plays a role, the Dutch government has embarked on a major reduction scheme. The goal is to reduce all ammonia emissions by 50-70% in the year 2000. In order to achieve the goal, ammonia emissions are considered which emanate from land spreading slurries.

Slurry injection is a well established method to reduce ammonia emissions from land spread slurries. Research by Bosma et al. (1976) indicates a 90% reduction. However, injection is not commonly used in the Netherlands. This is due to the costs of the system and the logistics involved in the supply of slurry. Furthermore, slurry injectors are known to be heavy. This prohibits application to low-lying peat soils in the western part of the Netherlands and to heavy clay soils in the northern and central regions.

Due to the limitations on the application of slurry injectors, other reduction methods have been developed in due course. Although methods are preferred which can be applied on both arable land and grassland, in most cases a distinction has been made between these two main types of land use. In the following paragraphs, the 'state of the art' on reducing ammonia emissions from land-spread slurries and manures in the Netherlands will be discussed. This will be preceded by a few notes on the experimental procedures involved in the measurements of ammonia emissions.

2. METHODS FOR MEASURING AMMONIA EMISSIONS IN THE FIELD

As soon as slurry is exposed to air, ammonia exchange to the boundary layer starts to take place. The exchange rate is dependent upon the vapour pressure. In case of slurry being applied to land, wind constantly refreshes the air keeping ammonia concentration in the boundary layer low. As a result of turbulence induced by radiation, convection and other factors an ammonia profile is created by the wind flowing over a field dressed with slurry. Techniques for measuring this profile, the micro meteorological model, have been described by Denmead (1983). A practical adaptation of the model for ammonia measurements following the spreading of slurries has been developed by Pain and Klarenbeek (1988).

It should be noted that measurements based upon micro-meteorological techniques are liable to meteorological influences. It can be concluded from experiments by Pain et al. (1990) and Klarenbeek (unpublished results) that rain occurring shortly after spreading significantly reduces ammonia emissions. In order to exclude meteorological uncertainties, a system of small wind tunnels (Lockyer, 1984) is also used for studying emissions developing following the spreading of slurries and manures. Both methods have specific advantages. Tunnels are easy to handle, take up little space and are not influenced by rain. However, as a result of size, tunnels are hardly suitable for experiments with coarse distribution of slurry over the experimental plot. This is the case with experiments involving a cultivation techniques in order to reduce ammonia emissions e.g. ploughing and harrowing. In such experiments the outcome is likely to be influenced by the place of a tunnel within the plot. Micro-meteorological experiments on the other hand are the best alternative when information is required on the emission rates under existing environmental conditions. It should be borne in mind that this type of experimental set-up is dependent upon the availability of land, labour and a laboratory with efficient (automated) analysing equipment to deal the large amount of samples produced.

3. AMMONIA EMISSION AND DECAY

Research into the ammonia emission following the spreading of slurries and manures has been undertaken in a number of projects of various research institutes in the Netherlands. Some of the results are reported by Pain and Klarenbeek (1988), Bruins and Huijsmans (1989) and Bruins (1990). It has been established that emission rates of ammonia -or fluxes- decay as time goes on. Especially during the first hours following the spreading, the flux decreases rapidly. As a result of differences in atmospheric stability during day and night time, a diurnal fluctuation in the flux can be observed. A typical emission curve is plotted in Figure 1.

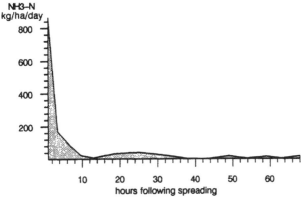

Figure 1. Ammonia emissions following land spreading of pig slurry.

Emissions from various slurries and manures in the Netherlands have been investigated by Bruins (1990) and Pain and Klarenbeek (1988) using tunnel and micro-meteorological methods. In all experiments pig and poultry slurry tends to produce higher ammonia emissions than cattle slurry (Table 1). This may be explained by differences in application rates and ammonium concentration of the slurry.

Table 1. Ammonia losses from slurries and manures.

Type of Application	Rate of Appl. (m^3/ha)	Nitrogen applied (NH_4-N) kg/ha)	Ammonia (NH_3-N) loss		
			Total (kg/ha)	% of HN_4-N	kg/m^{-3} Slurry
Cattle					
slurry 1	62	155	69.5	45	1.1
slurry 2	69	180	75.6	42	1.1
Pig					
slurry 1	39	134	44.1	33	1.1
slurry 2	35	89	38.5	43	1.1
slurry 3	34	86	44.8	52	1.5
slurry 4	33	84	44.3	53	1.3
slurry 5	26	54	44.0	82	1.7
slurry 6	62	260	84.9	33	1.4
Poultry					
broilers	7	50	10.0	20	1.4
air dried droppings	6.5	56	9.7	17	1.5
slurry from closed storage	30	241	90.5	38	3.0
slurry from open storage	47	254	74.7	30	1.6

Ammonia emissions from various rates of slurry application have been recorded by Pain and Klarenbeek (1988). The results are listed in Table 2. It can be concluded that emissions do not vary proportionally with the rate of application. This can be explained by the fact that higher rates tend to produce a bigger slurry layer has a higher resistance to evaporation. As a result of this, the release of ammonia into the atmosphere slows down.

Table 2. Ammonia loss from slurry at various application rates

Type of Application	Rate of Appl. (m³/ha)	Nitrogen applied (NH₄-N kg/ha)	Ammonia (NH₃-N) loss		
			Total (kg/ha)	% of NH₄-N	kg/m⁻³. slurry
Pig slurry (low rate)	10	43	22.9	53.0	2.3
Pig slurry (medium rate)	30	129	73.4	56.9	2.5
Pig slurry (high rate)	90	387	139.6	36.1	1.6

4. REDUCING AMMONIA EMISSIONS FROM LAND SPREADING

The goal for ammonia abatement set by the Dutch government has resulted in a number of experiments concerning the reduction of ammonia emission following land spreading. In most cases the possibilities of reduction have been investigated with respect to the type of land use e.g. arable land or pasture. Only a minor number of experiments were dedicated to techniques that are applicable in both cases. In all experiments, methods investigated were directly available to practical farming. Machines and methods being developed have not been considered.

4.1. General methods

There is a difference in ammonia emissions between day and night as can be derived from Figure 1. As stated, this is due to day and night variations in atmospheric. It is a tempting thought to use diurnal variations as means to reduce the emission. In order to check the possibilities of the method, two experiments have been conducted. During the first experiment (Fig. 2) extraordinary weather conditions existed. Temperatures were approx. 30 °C in the daytime and 27 °C at night. Furthermore, an open sky with no clouds and many hours of intensive sunshine were typical of the experimental period. In the second period (Fig. 3) the weather conditions were back to normal. From the results listed in table 3, it can be concluded that under normal atmospheric conditions land spreading of slurry at dawn reduces emissions considerably. No significant differences were observed for slurry spreading in the morning or the afternoon. This can be explained by the fact that most of the emission takes place during the first few hours following spreading. However, high temperatures combined with strong solar radiation during the first experimental period appeared have a special effects on the emission pattern. It caused a fast drying process of the slurry in the field. Slurry was observed to dry to a crust within 6-7 hours after being spread at 13:00 hours. In such cases little emission is expected. For evening spreading it is supposed that absence of solar radiation slows down the drying process, therefore leaving the slurry moist until the morning. As the atmospheric conditions were the same as for the preceeding day, all potential ammonia was evaporated in the daytime. The prolonged emission time compared to the other two times of spreading resulted in a higher total ammonia emission.

Table 3. Ammonia emission from land spreading slurry at different times.

Type of Application	Rate of Appl. (m^3/ha)	Nitrogen applied (NH_4-N kg/ha)	Ammonia (NH_3-N) loss		
			Total (kg/ha)	% of NH_4-N	kg/m^{-3} slurry
Cattle slurry at 9:00h	16	46	37	80	2.3
Cattle slurry at 13:00h	17	47	38	81	2.2
Cattle slurry at 20:00h	16	46	49	100	3.0
Cattle slurry at 9:00h	11	30	15	49	1.3
Cattle slurry at 13:00h	10	28	15	53	1.5
Cattle slurry at 20:00h	9	27	6	23	0.67

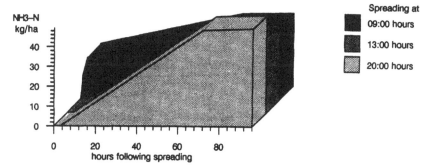

Figure 2. Ammonia emission following the land spreading of cattle slurry at different times; experiment 1.

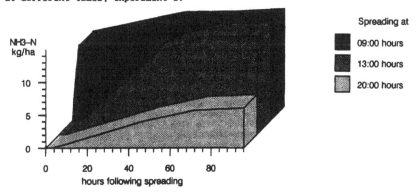

Figure 3. Ammonia emission following the land spreading of cattle slurry at different times; experiment 2.

As a result of the legislation on land application of animal slurry existing in the Netherlands, a certain quantity cannot be used for land spreading. The surplus has to be processed. At present two processing and drying plants are in operation. The product is sold to farmers as pellets. It is claimed that slurry pellets can be applied to grassland as well as to arable land with only a limited ammonia emission. In order to investigate the emission resulting from the application of slurry pellets an experiment was designed. In the experiment pellets were compared to pig slurry and a commercial ammonium fertilizer. Experimentation was carried in the autumn on arable with a shallowly ploughed wheat stubble. The weather conditions were stable with dew in the early in the morning hours. The results, listed in Table 4, clearly show a distinct reduction in the emission for the application of slurry pellets. The emission level approximates that of commercial fertilizer. It was observed that the ammonia emission from pellets started after dew had occurred. This suggests that moisture increases the ammonia flux from the pellets.

Table 4. Ammonia emission from slurry, fertilizer and slurry pellets.

Type of Application	Rate of Appl. (10^3 kg/ha)	Nitrogen applied (NH_4-N) kg/ha)	Ammonia (NH_3-N) loss		
			Total (kg/ha)	% of NH_4-N	kg.ton^{-1} product
Slurry pellets	2	49	8	16	4
Fertilizer	0.7	93	9	10	13
Pig slurry	27	171	113	66	4

4.2. Methods for reduction of emission on arable land.

As far as arable land is concerned, reduction of ammonia emissions can be achieved by covering the slurry with soil or by thorough mixing. Covering of slurry with soil can be attained by ploughing. Mixing is done by means of a rotary harrow or rigid tines. Since some of the arable land in the Netherlands is situated om mixed farms, the question is relevant whether incorporation of slurry can be delayed till the milking of cows has been finished. In order to obtain an answer to the question, four experiment were conducted. Apart from the weather conditions which were variable during the 4-week experimental period, the time elapsed between spreading and incorporation was taken as parameter in each experiment. The procedures applied and results are reported by Pain et al. (1990). A summary is listed in Tabel 5. Although experimental conditions varied, the relative effectiveness of the three incorporation methods was consistent throughout the experiment. Ploughing was most effective followed by the rotary harrow and the rigid tines. In order to achieve a 50% reduction of ammonia emission compared to that of the control plots, ploughing could be delayed until 6 hours after spreading while a rotary harrow can be used up to 3 hours after spreading. In all cases rigid tines did not reduce the emission satisfactory. The poor result may be explained by the fact that clods are easily produced during the operation.

Table 5. Ammonia emissions following incorporation at different times.

Incorporation Method	Time (h)	Rate of Appl. (m^3/ha)	Nitrogen applied (NH_4-N) kg/ha	Ammonia (NH_3-N) loss		
				Total (kg/ha)	% of NH_4-N	kg/m^{-3} slurry
Plough	0	64	500	20.6	5.9	0.32
Rotary harrow	0	64	500	45.7	12.9	0.70
Rigid tines	0	64	500	124.3	35.0	1.93
Control	0	64	500	206.7	58.2	3.21
Plough	3	62	590	27.2	7.4	0.45
Plough*)	3	64	436	29.0	10.5	0.47
Rotary harrow	3	62	590	38.4	10.3	0.62
Rotary harrow*)	3	64	436	57.5	20.2	0.90
Rigid tines	3	62	590	56.9	15.4	0.92
Rigid tines*)	3	64	436	93.1	30.4	1.36
Control	3	62	590	84.7	23.0	1.37
Control*)	3	64	436	139.2	48.5	2.17
Plough	6	61	527	46.9	13.5	0.78
Rotary harrow	6	61	527	46.9	16.7	0.96
Rigid tines	6	61	527	67.7	19.3	1.11
Control	6	61	527	101.9	29.1	1.67

*) Duplicated experiment.

4.3. Methods for emission reduction on grassland.

Since grassland has a more permanent character than arable land, methods for reduction of ammonia emissions following spreading differ. For pasture, it is of paramount importance that the sod is left undamaged. If this can not be avoided as is the case with injection, recovery should take place soon afterwards. It can be concluded from research results presented by Pain et al. (1989) that apllication of diluted slurry is a possible solution. The dilution ratio is mainly set by practical considerations. One part of slurry to three parts of water (1:3) is a generally accepted figure in the Netherlands. However, it is considered a negative factor that an extra amount of water has to be transported to the field in case of dilution. For this reason natural rain and irrigation of slurry following the land spreading is an alternative. Especially in the northern and western parts of the Netherlands where water can be obtained from ditches or shallow boreholes, irrigation is quite popular. An application rate of 10 mm of water is normal in this case.

Besides dilution and irrigation, deep injection and shallow sod injection are alternatives which are also applicable to pastures under normal meteorological conditions. In order to determine the effectiveness of these methods, a number of experiments have been undertaken. The results are listed in Table 6 and show a distinct effect of the latter method. This finding is along the same lines as those reported in the preceding paragraph. The highest reduction is obtained when slurries are quickly incorporated into the soil as is the case with injection. Nevertheless, irrigation and dilution produce acceptable reductions too.

Table 6. Ammonia emission following the spreading of slurry on grassland with different application methods.

Type of Application	Rate of Appl. (m^3/ha)	Nitrogen applied (NH_4-N kg/ha)	Ammonia (NH_3-N) loss		
			Total (kg/ha)	% of NH_4-N	kg/m^{-3} slurry
Deep injection					
(pig slurry)	40	223	0.45	0.2	0.01
(cattle slurry)	40	120	0.48	0.4	0.01
Sod injection					
(pig slurry)	20	112	4.48	4	0.22
(cattle slurry)	20	61	4.88	8	0.24
Irrigation (10 mm)					
(pig slurry)	10	57	9.12	16	0.91
(cattle slurry)	10	30	3.90	13	0.39
Dilution 1:3					
(pig slurry)	40	56	14.00	25	0.35
(cattle slurry)	40	38	7.22	19	0.18
Control					
(pig slurry)	10	57	27.36	48	2.74
(cattle slurry)	10	30	19.50	65	1.95

5. CONCLUSIONS

In view of the facts presented in the preceeding paragrahps, the following conclusion are relevant. Emissions during land application of animal slurries depent upon;

-The type of slurry. Pig slurry and poultry slurry produces higher emission rates than cattle slurry. Futhermore, slurry with a low dry matter content e.g. diluted slurry is known to have a low emission.
-The rate of application. A high application rate produces less ammonia m^3 of slurry being spread. Although the ammonia emission is not proportional to the application rate, the higher the rate the more ammonia being released to atmosphere.
-The machinery used for spreading. Emission declines as a result of the use of special applacation equipment like deep injection and shallow sod injection.
-The time of land spreading slurry e.g. morming or early evening affects the total ammonia emission. It should be noted that extreme weather conditions may reverse the result.

6. REFERENCES

Bosma, A.H., Klarenbeek, J.V., Krause, R. and Zach, M. 1976. Aparatus and Sytems for Odourless Land Application of Slurry (in Dutch). Publication 67. IMAG, Wageningen.

Denmead, O.T., 1983. Micrometeorological Methods for Measuring Gaseous Losses of Nitrogen in the Field. In: Gaseous Losses of Nitrogen fron Plant-Soil Systems (Freney & Simpson eds.) Martinus Nijhoff, The Hague.

Bruins, M.A., 1990. Odour and Ammonia Emissions during and following Land Application of Slurries (in Dutch), Report in preparation, IMAG, Wageningen.

Bruins, M.A., and Huijsmans, J.F.M., 1989. Reduction of Ammonia Emission from Pig Slurry following Land Application (in Dutch). Report 225. IMAG, Wageningen.

Pain, B.F., and Klarenbeek, J.V., 1988. Anglo-Dutch experiments on odour and ammonia. Research Report 88-2, IMAG, Wageningen.

Pain, B.F., Phillips, V.R., Clarkson, C.R. and Klarenbeek, J.V., 1989. Losses of Nitrogen through Ammonia Volatilisation during and following the Application of Pig or Cattle Slurry to Grassland. J Sci Food Agric, 47, pp. 1-12.

Pain, B.F., Phillips, V.R., Huijsmans, J.F.M. and Klarenbeek, J.V., 1990. Anglo-Dutch Experiments on Odour and ammonia Emission following the Spreading of Piggery Wastes on Arable Land. Research Report in preparation. IMAG, Wageningen.

Lockyer, D.R., 1984. A system for the Measurement of Losses of Ammonia through volatilisation. J Sci Food Agric, 35, pp. 837-848.

AMMONIA EMISSION AND CONTROL AFTER LAND SPREADING LIVESTOCK WASTE

K. Vlassak, H. Bomans, R. Van den Abbeel
K.U. Leuven, Faculty of Agricultural Sciences
Laboratory of Soil Fertility and Soil Biology
Kardinaal Mercierlaan 92, B-3030 Leuven

Abstract

This paper deals with the ammonia volatilization from arable land.
Volatilization from liquid pig manure was measured using a modified
windtunnel method. Up to 40 % of the total ammonium nitrogen content of
the slurry was lost as NH_3 during the first 4 days following application.
Fifty per cent of the NH_3-gas was lost during the first twelve hours and
70 % was volatilized after 24 hours following application.
Reduction of the NH_3 losses was studied using chemical substances.
A reduction of 50 to 90 % was obtained.

I. INTRODUCTION

Since many years it has been known that large losses of nitrogen occur
from livestock wastes and chemical fertilizers.
Also, it has been recognized that ammonia volatilization is a significant
pathway of loss of nitrogen, that ammonia is lost rapidly after spreading
on land, and that a substantial part of the NH_3 is deposited on nearby
sites and can contribute to soil acidification upon transformation into
HNO_3 through nitrification.
The total ammonia emission in Europe is estimated to be 6.4 Mt of NH_3 per
year with a major contribution of above 80 % from livestock wastes.
According to Buijsman (1987), the highest emission densities in Europe
are to be found in the Netherlands (6.4 t km^{-2} yr^{-1}) and in Belgium
(5.3 t km^{-2} yr^{-1}).
Indeed the highest concentration of animals is found in the Netherlands
and Belgium as indicated in Table 1. This data from the databank Regio
EEC shows that very high densities occur in some areas in both of the
countries. In most of the cases the expansion of the animal livestock has
taken place on farms with limited land surface and here problems arise in
relation to an appropiate use of the animal manure.

TABLE 1 AREAS OF THE EUROPEAN COMMUNITY WITH THE HIGHEST DENSITY OF
LIVESTOCK (CATTLE, PIGS), 1987

AREA	COUNTRY	TOTAL LU 100 ha AA
1. Noord-Brabant	NL	706.82
2. Gelderland	NL	571.89
3. West-Vlaanderen	B	543.91
4. Limburg	NL	518.76
5. Antwerpen	B	511.56
6. Overijssel	NL	473.18
7. Utrecht	NL	428.96
8. Oost-Vlaanderen	B	404.07
9. Limburg	B	309.35
10. Muenster	D	284.84

Problems concerning nitrate content in drinking water and eutrofication of the surface water are very common. Besides NH_3 volatilization, N_2O production through dinitrification is in certain conditions an urgent problem. This paper summarizes current research on ammonia volatilization using windtunnels in the measurements of ammonia loss, following the application of livestock slurries to land and discuss chemical methods for reducing losses of ammonia by volatilization.

2. MATERIALS AND METHODS

Site characteristics

Measurements of NH_3 volatilization were carried out on a fallow plot at the university experimental site. The soil was loamy with a Bt-horizon (Typic Hapludalf, USDA). The experiments on NH_3 volatilization control were carried out on a fallow sandy soil, (Orthod, USDA).
Selected soil properties measured before adding manure are given in table 2. Soil and air temperature, rainfall, wind direction and wind speed were measured continuously at the university experimental site.

TABLE 2   SOIL CHARACTERISTICS

Soils	pH(H_2O)	pH(KCl)	TOTAL C %	TOTAL N %	NO_3^--N Kg/ha	NH_4^+-N Kg/ha
Typic Hapludalf	7.1	5.9	0.63	0.07	10.0	4.3
Orthod	6.5	5.4		0.24	0.5	2.0

Slurry treatment

For the NH_3 volatilization, pig slurry was applied to the soil surface in two dressings (winter and spring) at a rate of 50 t ha^{-1}. For NH_3 volatilization reduction studies 50 t ha^{-1} of cattle, pig or chicken manure was used. The characteristics of the slurries are found in table 3.

TABLE 3   CHARACTERISTICS OF THE SLURRIES USED IN THE EXPERIMENTS

	pH	DM%	OM%	Tot. N%	$NH_4^+-N%$
pig slurry					
- winter dressing	6.6	4.4	-	0.54	0.42
- spring dressing	7.1	6.1	-	0.50	0.35
pig slurry	6.7	3.7	2.5	0.40	0.28
cattle slurry	6.9	8.4	6.6	0.38	0.18
poultry slurry	7.2	3.7	2.2	0.50	0.43

Ammonia volatilization measurements.

The NH_3 losses were measured using aerodynamically shaped soil covers in which a constant air flow could be directed over a specified surface area of 500 cm^2, as described by Van den Abbeel et al., 1989, a constant flowrate, 17.5 volumes min^{-1}, was maintained through the losses by means of a vacuum cleaner.

According to several authors (Watkins et al., 1972; Hargrove et al. 1977;
Fenn and Kissel, 1973; Rachpal-Singh and Nye, 1986) the limiting factor
for NH_3 volatilization is the transfer coefficient rather than the wind
speed.
The NH_3 in the air stream was trapped in 50 ml boric acid solution. Two
NH_3 scrubbers were used in series in order to reduce losses through
evaporation and aerosol formation. NH_3 was determined afterwards by
tritrating the boric acid solution with H_2SO_4 0.01 N.

3. RESULTS AND DISCUSSION

Ammonia volatilization experiments

The main total losses recorded in the modified windtunnel experiments for
spring and winter treatments are presented in Fig. 1.

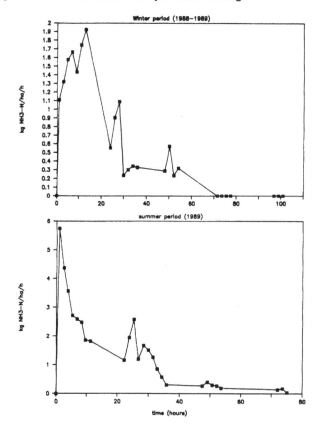

Fig. 1. Ammonia volatilization in kg NH_3-N ha^{-1} h^{-1} following application
in summer and winter period

For both dressings the same pattern of losses is observed. Within a few
hours after slurry application the ammonia volatilization rate increased
to a maximum level of 5.8 kg N ha^{-1} in spring and 1.9 kg N h^{-1} ha^{-1} in
winter treatments, followed by a rapid decrease. For both slurry
dressings a small peak was observed after 24 hours and after 48 hours of
application. This small increase corresponds with an increase in
temperature during daytime. After 3 days only negligible amount of NH$_3$
losses were recorded.

Total losses following application of the wastes to land are presented in
Fig 2. Pig slurry applied in spring time shows a very pronounced loss of
NH$_3$. A loss of 78 kg NH$_3$-N was recorded. It means that almost 40 % of the
applied NH$_3$-N is volatilized. In winter period only 48 kg NH$_3$-N loss was
registrated which corresponds with 23 % of the added NH$_4^+$-N.

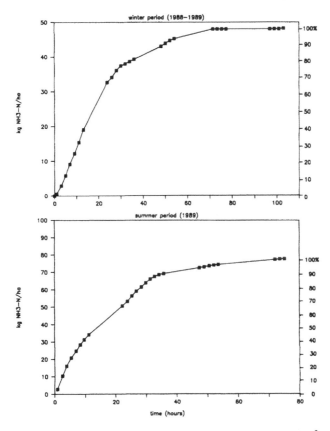

Fig. 2 Cumulative volatilization of ammonia in kg NH$_3$-N ha^{-1} follwing
application in summer and winter period.

It can also be seen from Fig. 2 that 12 hours and 24 hours following the
application, 50 and 70 % of the ammonia is volatilized respectively.
This indicates that incorporation in the soil to reduce ammonia
volatilization, has to be carried out as soon as possible after
application.
The results confirm earlier measurements where rates of NH3-N losses of
3.5-4 kg N h^{-1}ha^{-1} or 60 % of the NH$_4$$^+$-N were recorded (Van den Abbeel et
al., 1989). These losses were within the range of values reported by
Thompson et al. in 1987, Pain 1988, Döhler and Wiechmann 1987, Lauer et
al. 1976, Van der Molen et al. 1989. According to several authors most of
the NH$_3$ is lost very shortly after application. The total loss that
Thompson et al. 1987 recorded after surface application of slurry in
spring time was 48 % of the initial NH$_4$$^+$-N . Variation in measurements of
NH$_3$ volatilization rates are due to many parameters influencing NH$_3$
volatilization such as soil type, weather conditions, slurry
characteristics and management practice.

Reducing losses of ammonia
An attempt was made to reduce losses of ammonia using chemical
components. In a first experiment CaCl$_2$ (33% W/V), MgCl$_2$ (47% W/V) and
formaldehyde (40% V/V) were used to study the effect on NH$_3$
volatilization.
The results of these experiments are given in Fig. 3 for CaCl$_2$ and MgCl$_2$
and Fig. 4 for formaldehyde.

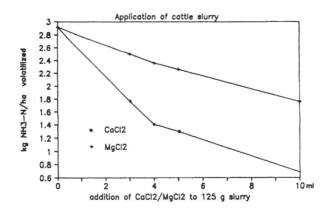

Fig. 3. Reduction of ammonia volatilization by use of MgCl$_2$ and CaCl$_2$
solution.

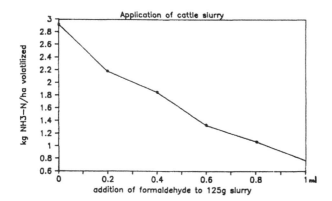

Fig. 4. Reduction of ammonia volatilization by use of formaldehyde additions in ml per 125 g slurry.

From this figure it can be seen that each of the components has a reducing effect on the rate of NH₃ volatilization.
CaCl₂ and formaldehyde have a pronounced effect with increasing concentrations.
But the amounts to add were far too large to obtain a strong positive effect.
Experiments were carried out with different combinations of these elements. A carefully chosen combination existing of a mixure of 1/3 of the CaCl₂ solution , 1/3 of the MgCl₂ solution and 1/3 formaldehyde given in a concentration of 1.5% to the slurry has given extremely good results as can be seen from Fig. 5a, 5b, and 5c for cattle, pig and chicken slurry respectively.
The NH₃ emission pattern is identical for the treated and untreated manure. The total NH₃ emission amounts respectively 58 and 57 % for cattle and pig slurry after 36 hours and 42 % for poultry manure after 28 hours. In the case of poultry manure the NH₃ losses are still increasing after 28 hours.
The total ammonia loss is reduced 70, 85 and 44 % respectively for cattle, pig and chicken manure as can be seen from Fig. 6a, 6b and 6c.
Other experiments demonstrated the potential for reducing ammonia loss by 50 to 60 % by acidifying the slurry with sulphuric acid prior to surface application or by injection the slurry (Pain et al.,1987). But in both the cases reducing loss of ammonia increased losses of N through denitrification.

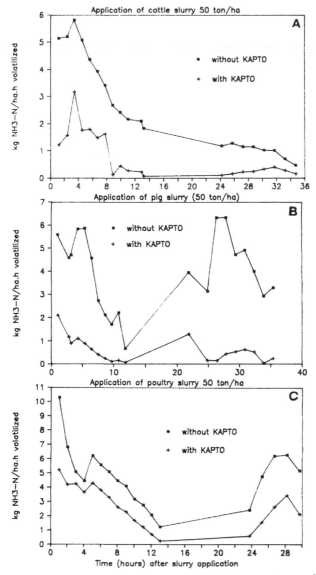

Fig. 5 Reduction of ammonia volatilization rate in kg NH_4^+-N ha^{-1} h^{-1} following application of cattle slurry, pig slurry, chicken slurry in 50 ton ha^{-1}.

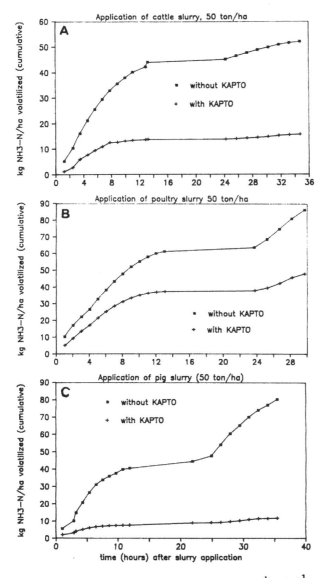

Fig. 6 Cumulation volatilization of ammonia in kg NH_4^+-N ha^{-1} following application of cattle slurry, pig slurry and chicken slurry in 50 ton ha^{-1}.

4. CONCLUSIONS

From this study it can be concluded that most of the slurry nitrogen volatilizes directly as NH_3 when applied to the soil. The results show that in summertime the NH_3 emission was 40 % and in wintertime 23 % of the NH_4 present in the slurry.
The volatilization of NH_3 can be reduced with 50-90 percent by adding additives. Further research is required to know the behaviour of the treated slurry following application to the soil and to know the way ammonia is immobilized.

5. ACKNOWLEDGEMENTS

We gratefully acknowledge the financial support from IWONL and the technical assistance from C. De Ruysscher, K. Cassaert and C. Jordens.

REFERENCES

Buijsman E., 1987. Ammonia emission calculation fiction and reality. In proceedings of the Ammonia and Acidification Eurasap Symposium, Asman W.A.N. and Diederen S.M.A. (eds), pp 13-27. Netherlands, RIVM, TNO.

Van den Abbeel R., Claes A., and Vlassak K., 1989.
Gaseous Nitrogen Losses from Slurry-manured land.
In Nitrogen in Organic wastes applied to Soils.
Jens Hansen and Kaj Henriksen eds., pp. 213-224, Academic Press.
Harcourt Brace Jovanovich, Publishers London.

Watkins S.H., Strand R.F., De Bell D.S. and Esch J., 1972.
Factor influencing ammonia losses from area applied to northwestern forest soils. Soil Science Society of America Journal, 36 , 334-357.

Hargrove W.L., Kissel D.E., and Fenn L.B., 1977. Field measurements of volatilization from surface applications of ammonium salts to calcereous soils. Agronomy Journal, 69, 473-476;

Fenn L. B., and Kissel D. E., 1973. Ammonia volatilization from surface applications of ammonium compounds on calcereous soil.
1. General theory, Soil Science Society of America Proceeding, 37, 855-859.

Rachpal-Singh, and Nye P.H., 1986, A model of ammonia volatilization from applied area. 2. Experimental testing. Journal of Soil Science, 37, 21-29.

Thompson R.B., Ryden J.C. and Lockyer D.R., 1987. Fate of nitrogen in cattle slurry following surface application or injection to grassland. Journal of Soil Science, 38, 689-700.

Pain B.F., 1988. Ammonia losses during and following the application of slurry to land. In safe and Efficient slurry utilisation, COST - FAO workshop Liebefeld/Bern in press.

Döhler H., and Wiechmann M. 1988. Ammonia volatilization from liquid manure after application in the field. In agricultural waste management and Environmental Protection. In proceedings of the 4th International Symposium of CIEC, Braunschweg, FRG, Welte E., and Szabole I. eds. Goettingen, Goltze-Druck.

Van der Molen J., Bussink D.W., Vertregt N., Van Faassen M.G., and Den Boer D.J. 1989. Ammonia volatilization from arable and grassland soils. In "Nitrogen in organic wasted applied to soils. Jens Hansen and Kaj Henriksen eds. pp 185-201, Academic Press , Harcourt Brace Jovanovich, Publisher London.

Lauer D.A., Bouldin D.R., and Klaussner S.D. 1976. Ammonia volatilization from dairy manure spread on the soil surface. Journal of Environmental Quality, 5, 134-141

Pain B.F., Thompson R.B., de la Lande Cremer L.C.N., and Ten Holte L., 1987. The use of additives in livestock slurries to improve their flow properties, conserve nitrogen and reduce odours. In Animal Manure on grassland and Fedder Crops Van der Meer (ed), Dordrecht : Martinus Nijhoff.

AMMONIA EMISSIONS DURING AND AFTER LAND SPREADING OF SLURRY

A. AMBERGER

Institute of Plant Nutrition,
TU München–Weihenstephan, D–8050 Freising 12

Summary

Ammonia emissions during and after land spreading of slurry can rise up to 80% of the applied NH_4–N and occur mainly in the first hours after spreading. Mode of application, slurry dry matter content, atmospheric temperature and soil moisture are the essential hazard factors. Application of highly viscous slurry on stubbles and straw, on compacted soils or grassland impairs infiltration into the soil and results in high ammonia volatilization. Immediate incorporation into the soil, application at low temperature and possibly diluting of viscous slurry are proper measures to decrease ammonia losses and air pollution.

1. INTRODUCTION

Ammonia emissions into the atmosphere are supposed to be about 80% and more due to animal production. During and after land spreading of animal slurry they can appear in considerable or even very high quantities. On the one hand they give rise to environmental – toxicological problems concerning forest decline and acidification of soils, on the other hand economically they are real losses of nitrogen to the farmer (1,2,3,6).

Basically the pattern of ammonia release shows a logarithmic curve increasing rapidly with pH and temperature. Under practical conditions the main parameters are: temperature and flow of air, soil properties, vegetation, type (dry matter content) and mode of application of slurry.

2. MATERIAL AND METHODS

The common methods to determine ammonia losses are:

a) closed dynamic system under controlled conditions (growth chamber) (5)
b) wind tunnel under field conditions (4)
c) micrometerological measure system
d) indirect and unspecific method by different N–uptake in plant experiments

In our institute we have worked with the method a) and b) (air flow rate equivalent to a 35–fold air exchange/min), sometimes completed with d). Slurry was applied on the basis of equal amounts of ammonium–N (commonly used in the practice):

– 100 kg NH_4–/ha on arable land
– 60 kg NH_4–/ha grassland

3. RESULTS AND DISCUSSION

Ammonia losses of slurry as influenced by mode of application, dry matter content of slurry and temperature.

Mode of slurry application is the most important factor; while losses were very high (50%) after surface application, incorporation with a cultivator reduced them to about one fifth (fig 1). Due to mixing of slurry and soil, more slurry ammonium is getting into contact with sorptive sites of the soil and thus escaped potential evaporation (4).

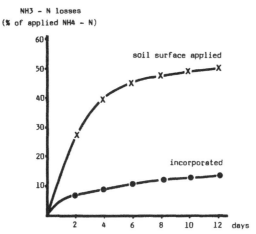

Fig. 1:   Ammonia losses from cattle slurry dependent on application mode (silty loam, pH 6.5)

In another experiment (5) under controlled conditions (fig.2) the greatest losses (about 40%) appeared already in the first hours after surface application and decreased later (accumulated curve). Ammonia volatilization rised with temperature (5°C to 20°C) and dry matter content (1 – 6.4–8%) of slurry.

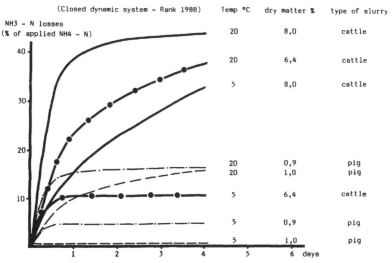

Fig. 2:   Ammonia losses from slurry depending on dry matter content and temperature

128

Application of slurry with 10% dry matter on stubbles and wheat straw (4) in August (28°C) reached losses of 70% compared with 40% in case of 5% dry matter (fig. 3). After November application (14°C) on a compacted soil after corn harvest the losses decreased from 40% to 10% (with 7.8 resp. 1.5 % dry matter). In both cases the effect is due to differing infiltration: slurry with low viscosity infiltrated into the soil more rapidly and thus enabled a stronger sorption of NH_4-N. Highly viscous slurry remained on the soil surface for a longer period with direct contact to the atmosphere.

Compared with these factors the pH of the slurries (generally between 6.8 and 8.0) is not so relevant, and therefore not described here in details.

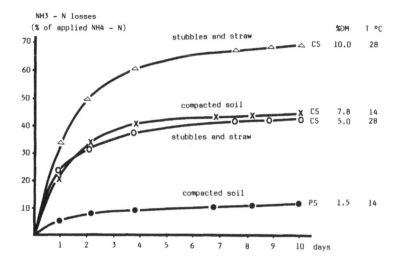

Fig. 3:   Ammonia losses after surface application of cattle (CS) and pig slurry (PS) on stubbles and straw (Aug.) or compacted soil (Nov.) on silty loam (pH 6.5).

Ammonia losses as influenced by soil type and moisture

Concerning the type of soil (5) highest losses were found on the light sandy soil, lowest on silty loam again with great differences between surface applications and incorporation (fig. 4). With respect to soil moisture (fig. 5), in a dry soil (10% of total water capacity) the losses are considerably higher (5) than in a soil with 30% of total water capacity).

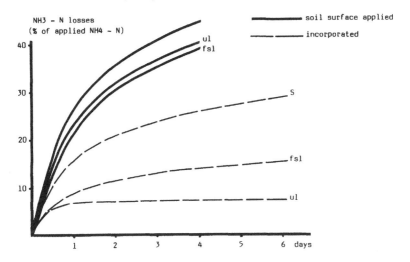

Slurry: 50mg NH N/pot

NH3 – N losses
(% of applied NH4 – N)

soil surface applied

incorporated

Fig. 4: Ammonia losses from slurry on silty loam (pH 6.5) fine sand loam (pH 7.3) and sand (pH 6.0) Temp. 20°C

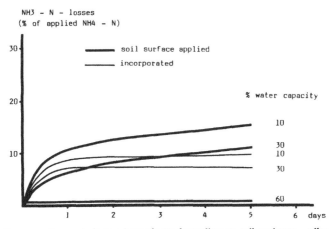

Slurry: 50 mg NH4 – N/pot

NH3 – N – losses
(% of applied NH4 – N)

soil surface applied

incorporated

% water capacity

Fig. 5: Ammonia losses from slurry depending on soil moisture – silty loam, pH 6.5, 20°C

Slurry application to growing crops

One of the possibilities to use slurry in spring is the application to growing crops (4) in lower quantities (20 – 30 m³/ha). A crop canopy with winter cereals (fig. 6) decreased ammonia volatilization by 50 % and more against surface application on a bare soil on behalf of lowering wind velocity close to the soil surface and partly also direct uptake of NH₃ by plants.

silty loam pH 6.5 (Windtunnel - Huber 1989)

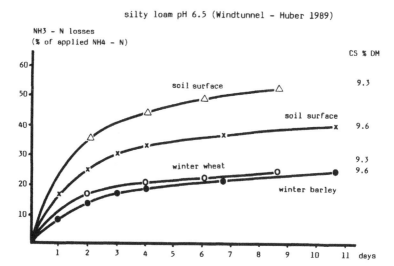

Fig. 6:   NH₃ losses after spring application of cattle slurry (CS) into growing crops (winter wheat, winter barley: EC 29–32) or on soil surface resp.

Slurry application to grassland

On grassland extremely high losses (nearly 80 %) were observed after slurry application in late summer with dry and warm weather and a high dry matter content of the slurry (fig. 7). With low temperatures in winter (minimum –2°C, maximum 15°C) losses resulted to 20 % and were not very much higher than after surface application on bare land. Obviously the sod impaired the infiltration of slurry, so that less NG₄₊ could be sorbed by the soil. In this case slurry diluted with water can reduce the losses greatly.

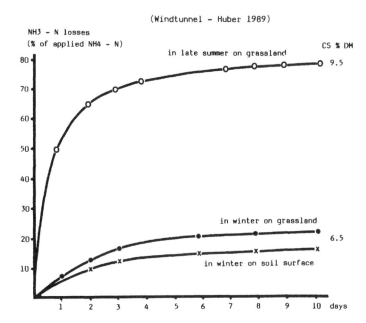

Fig. 7:   NH₃ losses from cattle slurry (CS) on grassland – silty loam pH 6.5

REFERENCES
(1)   AMBERGER, A. (1987). Utilization of organic wastes and its environmental implications. Proceedings of 4th Int. CIEC Symposium 1987, Braunschweig-Völkenrode 2, (315 – 320)

(2)   AMBERGER, A. (1989). NH₃-Verluste aus der Anwendung organischer und anorganischer Dünger. Kongreßband VDLUFA Bayreuth (im Druck)

(3)   AMBERGER, A. and HUBER, J. (1988). Ammonia losses after animal slurry application. In: Safe and efficient slurry application. Commission of the European Communities. Concerted action, hold in Liebefeld June 1988, (239 – 247)

(4)   HUBER, J. (1989). Versuche zur Quantifizierung verschiedener Einflußfaktoren auf Ammoniakverluste nach Gülledüngung. Dissertation Technische Universität München, 1990

(5)   RANK, M. (1988). Untersuchungen zur Ammoniakverflüchtigung nach Gülledügung. Dissertation Technische Universität München, 1988

(6)   RANK, M., HUBER, J. and AMBERGER, A. (1987). Mode trials on the volatilization of ammonia following to slurry application under controlled climate and field conditions. Proceedings of 4th Int. CIEC Symposium 1987, Braunschweig-Völkenrode

LABORATORY AND FIELD EXPERIMENTS FOR ESTIMATING AMMONIA LOSSES FROM PIG AND CATTLE SLURRY FOLLOWING APPLICATION

H. DÖHLER

University of Bayreuth
Dept. of Agroecology
P.O. Box 101251
D-8580 BAYREUTH (FRG)

SUMMARY

Field and laboratory experiments were conducted to determine ammonia losses from pig (PS) and cattle slurry (CS). Ammonia losses appeared to be closely related to temperature. Two days after application 80-90% of total losses were established.

In the laboratory study total NH_3-N losses of 28-60% (of applied ammonia N) from CS and 4-32% from PS were recorded. Even if slurry was applied at high temperatures, ammonia losses from PS were less by half than from CS. These differences are mainly due to the flow properties of pig and cattle slurry. In contrast, ammonia volatilization from a PS with high ammoniacal N, high dry matter content and high pH turned out to be approximately as high as from CS. Total losses from PS were nearly twice as high, when slurry was applied to a compacted soil surface.

In field experiments ammonia losses from surface applied slurry were highest (54-67%) if slurry was applied to soils covered with crops or plant residues and if temperatures were high. Volatile losses were smallest (3%/23%), if PS and CS were applied to a well structured, slightly frozen soil.

Incorporation and injection of CS turned out to be very effective in reducing ammonia volatilization (1-5%). Also slurry application by a tow hose spreader compared with application by a baffle plate spreader reduced losses by nearly one third.

The effects of slurry treatments -dilution with water, separation, and addition of acids- on reducing ammonia volatilization are discussed.

INTRODUCTION

Already at the beginning of the 19th century variable yields have been assumed to be due to ammonia losses after manuring livestock wastes and therefore it has been recommended to apply liquid manure during raining periods (1). Meanwhile, it has been documented by many investigations, that recoveries of N from surface applied slurries are low and variable (2,3,4). This is mainly due to ammonia volatilization. Consequently, there is often some uncertainty, when farmers have to calculate the fertilizer value of their slurries and therefore, slurry is treated as a waste product and not as a compound fertilizer.

Additionally, concern has grown in the last years over ammonia as an environmental pollutant. Ammonia which is lost from slurries not only represents a decrease in fertilizer value, but also is a source of air pollution. Especially forest ecosystems are said to be susceptible to high ammonia and ammonium deposition (5,6,7). The objectives of the present study were to determine the major factors influencing ammonia losses by laboratory experiments and to quantify ammonia losses from pig and cattle slurries by field experiments.

MATERIALS AND METHODS

Laboratory experiments
Experiments were performed using a closed, dynamic system. The NH_3 collection system consisted of four components: one scrubber containing water for air moistening, an acid trap containing a 2% boric acid solution for collecting NH_3 gas and a volatilization chamber where slurry was applied to soil. Volatilized ammonia gas was pulled into the acid trap by a vaccum pump.

The slurries used in the experiments were collected from the slurry storages of housed dairy cattle (CS I/CS II), from a pig breeding (PS I) and a pig fattening stock (PS II). The composition of slurries, and some properties of the soils used in the laboratory experiments are given in tables 1 and 2. The quantity applied corresponded to 35 m^3ha^{-1}. Composition of slurries used in the field study were similar to those used in the laboratory study.

	pH	DM (%)	NH_4-N (kg m^{-3})	N_t (kg m^{-3})
CS I	8,1	8,0	2,2	3,9
CS II	7,6	9,6	1,5	3,3
PS I	7,5	3,9	3,5	5,3
PS II	8,0	7,9	5,9	7,7

CS = CATTLE SLURRY   PS = PIG SLURRY

Table 1.   Composition of the slurries used in the laboratory experiments

	pH ($CaCl_2$)	C_{org} (%)	N_t (%)	particle size distribution sand	silt (%)	clay	CEC (meq/100g)
loamy soil	7,5	3,2	0,28	14	55	31	32
sandy soil	5,7	1,8	0,15	63	24	13	16

Table 2.   Properties of the soils used in the laboratory experiments

Field experiments

Loss of N from slurries applied in the field through ammonia volatilization was determined using the mass balance method described by BEAUCHAMP et al. and RYDEN & MC NEILL (8,9). They supposed the vertically integrated product of wind speed and atmospheric NH_3 concentration to be equal to the NH_3 flux from the soil surface. Slurry was applied on a circular area with a radius of 30 m. For estimating ammonia flux wind speed and ammonia concentration were determined in the center of the area at five heights for a time of usually two days.

The field experiments were conducted in Memmelsdorf, Northern Bavaria, about 80 km north of Nuremberg. Soils were of diluvial type. Average properties of the soils were pH 6,0-6,5 ($CaCl_2$); sand 40-50%, silt 40%, clay 10-15%; total N 15%; organic carbon 1,5%; CEC 25 meq 100g^{-1}.

Slurry application rates were 20, 21, 35 and 40 m^{3}ha^{-1} for cattle slurry and 21 and 23 m^{3}ha^{-1} for pig slurry, respectively. As far as surface application is concerned, slurry was usually applied by a pump trailer tank with a baffle plate spreader. Additionally a slurry injector (three experiments) and a pump trailer tank with a tow hose spreader (one experiment) were used.

RESULTS AND DISCUSSION

Laboratory experiments

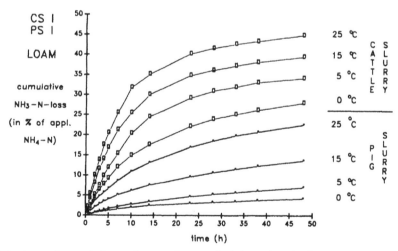

Fig. 1. Ammonia losses from surface applied pig and cattle slurry (PS I/CS I): Effect of temperature

Figure 1 presents the cumulative ammonia losses from pig and cattle slurry applied to the loamy soil at various temperatures. In all experiments the highest rates of ammonia losses occured during the initial hours after application and then decreased in an exponential manner. With increasing temperatures both ammonia loss rates and total ammonia losses increased. When temperature raised from 0 °C and 25 °C, losses from CS increased from 28% to 45%. Ammonia lost by volatilization from PS appeared to be considerably smaller at the same conditions. From this slurry only 4% and 25% were lost, respectively. These differences are due to the different slurry flow properties. While flow properties of PS are similar to those of water, CS has a by higher viscosity. Therefore PS infiltrates into the soil more rapidly and slurry ammonium ions are sorbed to the soil colloids. On the contrary viscous CS only infiltrates partially and ammonia volatilizes from the slurry liquid covering the soil surface.

Comparing various types of pig and cattle slurrry at temperatures of 15 °C, total losses from PS II of 27% were measured, while losses from PS I amounted only to 14% (table 3). This difference is attributed to the properties of PS II (high dry matter and ammonia content, high pH) which have already been documented to be important factors enhancing ammonia losses (10,11). Differences between cattle slurries were found to be relatively small.

slurry	PS I	PS II	CS I	CS II
ammonia N losses (% of applied NH_4-N)	14	27	39	34

Table 3.   Total ammonia losses from surface applied pig and cattle slurry 2 days after application at 15 °C: Effect of different slurry types

Table 4 shows ammonia volatilization from PS I, when applied to a sandy and a loamy soil or a plastic sheet. Apparently, hindering slurry infiltration into the soil by the plastic sheet nearly caused a total loss of initial slurry ammonium N. And, despite of the high pH (7,5) of the loamy soil ammonia losses were higher from the sandy soil (pH 5,7). Obviously the effect of soil reaction was compensated by the higher CEC of the loamy soil. Another laboratory study indicates (12), that this effect was less important if CS was applied.

	pig slurry applied to		
	loamy soil	sandy soil	plastic sheet
ammonia N losses (% of applied NH_4-N)	14	22	81

Table 4.   Total ammonia losses from surface applied PS I 2 days after application at 15 °C: Effect of soil type and a plastic sheet.

		ammonia losses	
slurry	properties of soil surface	2 days after application	4 days
CS	well structured	44	48
	compacted	53	60
PS	well structured	14	19
	compacted	22	32

Table 5.   Ammonia losses (% of applied NH_4-N) from surface applied CS and PS two and 4 days after application to the loamy soil at 15 °C: Effect of soil compaction

Soil structure is an important factor influencing ammonia losses (table 5). Especially losses from pig slurry increase considerably if slurry is applied to a compacted soil surface (ammonia loss: 32%) compared with application the to a well structured soil (19%). Through compaction of the soil surface slurry infiltrates less rapidly and therefore ammonia emission is favoured. As CS infiltrates only partially even into a well structured soil, the effect of soil compaction on total losses from CS is less important.

Results in table 5 also show that already two days after application 80-90% of total ammonia losses (4 days after application) were already volatilized. ?

Field experiments

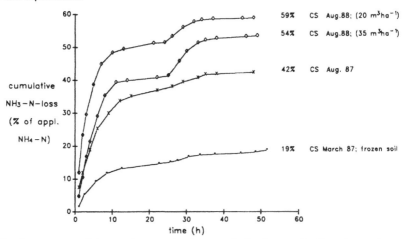

Fig. 2.   Ammonia losses from cattle slurry applied to bare soils. Experiments in August 1988 were performed simultaneously

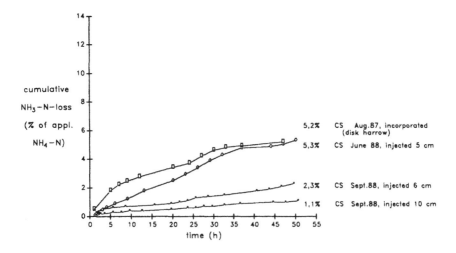

Fig. 3.   Ammonia losses from incorporated and injected cattle slurry. Experiments
in September 1988 were performed simultaneously

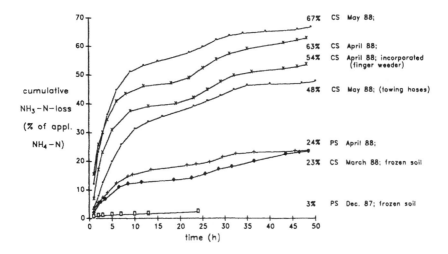

Fig. 4.   Ammonia losses from pig and cattle slurry applied to winter wheat in winter
and in spring. Experiments in April (CS) and May were performed
simultaneously

The results of field and laboratory experiments were similar. Cumulative ammonia loss curves from cattle slurry applied to the surface of bare soils are given in figure 2. If slurry was applied to a frozen soil, losses were relatively small (19%). Application at high temperatures caused higher losses of 42-59%. The rapid increase of loss rates on the second day in the August (1988) experiments were due to an increase of wind speed.

Losses from the cattle slurry with the low application rate (20 m^3ha^{-1}) were higher than from the high application rate (35 m^3ha^{-1}). This is probably due to the thinner slurry layer, which is rapidly passed by ammonia gas by diffusion. However, the differences between the application rates were relatively small. These results indicate a descrepancy between the present study and results described previously (13), where increasing ammonia losses were found as application rate increased.

Figure 3 shows that incorporation of CS into the soil gave smaller ammonia losses (5%) than surface application at similar weather conditions. Slurry was mixed to a depth of about 10 cm with a disc harrow right after application. This result is confirmed by the previous findings, indicating that ammonia losses from incorporated slurries are about one order of magnitude lower than from surface applied slurries (14). One can further see that slurry injection is the most effective way to reduce volatilization. In the September experiments losses turned out to be neglegibly low (1 and 2%) whereas in the June experiment losses of 5% were measured.

Since N losses by leaching of NO_3 following slurry application in autumn may be substantial, it has often been recommended to apply slurry to growing crops in spring. Therefore ammonia losses from slurries applied to winter wheat in winter 1987 and in spring 1988 were measured. Results are given in figure 4: Application of pig and cattle slurry to a slightly frozen, well structured soil which allowed rapid slurry infiltration, gave the smallest ammonia losses amounting of 3% (PS) and 23% (CS) of applied NH_4-N, respectively.

The application of PS to tillering winter wheat at mean daily temperatures of about 10 °C resulted in losses of 24%. In this experiment ammonia volatilization was favoured by the compacted soil surface and clogged soil pores caused by heavy rainfalls prior to the experiment.

In April and May 1988 two simultaneous experiments were conducted each. The distance between the experimental plots was about 200 m. During the experimental periods warm and mostly sunny weather prevailed. In the April experiments each plot was top dressed with CS (stage of growth: shooting). An attempt was made to reduce ammonia losses by working in the CS with a finger weeder. Because of the high temperatures, losses amounted to 63% from surface applied slurry. However, only initial loss rates were depressed by incorporation. The pattern of volatilization was similar to that of surface applied slurry, and differences between total ammonia losses from surface applied and incorporated slurry were small (63 and 54%, resp). The soil properties may explain, why incorporation was quite ineffective. Due to the warm and dry weather conditions the compacted soil was crusted, and therefore the working depth of the finger weeder had not been sufficient for intensive mixing of soil and slurry. Additionally, ammonia losses were favoured by the vegetation, because viscous CS partially adheres to the crop leaves. If the contact of CS with the soil was hindered completely by crop residues (chopped straw), a nearly total loss of applied ammonium N was found (14).

In the May experiments (stage of growth: last leaf) the effect of different surface application systems on ammonia emissions from CS were investigated. Losses from the cattle slurry applied with the baffle plate spreader (67%) were substantially higher than losses from the cattle slurry applied with a tow hose spreader (48%). The advantage of tow hose application is that slurry is applied directly to the soil surface and therefore not any slurry adheres to crop leaves. Also wind speed is substantially lower near the soil surface.

In most experiments losses mainly occur during the first 10 hours following application. In contrast to results found in other investigations (8,11), ammonia flux pattern did not appear to be closely related to air or soil temperature on the day of application. This was only true for the days following. On the day of application it has often been established, that with decreasing loss rates air and soil temperatures are increasing.

CONCLUSIONS

The manurial value of slurry application mainly depends on avoiding N losses. To get a sufficient manurial effect ammonia losses should not exceed 30% of applied ammonium N. From the results of the present study the conclusion is drawn, that pig slurry N is effective, if applied at temperatures less than 20 °C.

As shown above, ammonia losses from cattle slurry already amount to 30%, if applied at temperatures of less than 5 °C. Keeping losses below 30% is only possible by application of slurry to frozen soils. Since there is a risk of surface runoff, slurry should only be applied to slightly frozen soils, which are just capable of carrying spreading equipment. If temperatures exceed 5 °C, the farmer has to take measures to reduce losses.

Incorporation and injection of slurry into the soil turned out to be very effective. However, high losses by denitrification from injected slurries have been observed (15).

If incorporation of slurry is not possible ammonia losses can be reduced by application of CS with a tow hose spreader. However, this application technique is very susceptible to blockage by coarse matter.

Additionally, several slurry treatments have been shown be effective in reduction of ammonia emission:

In a laboratory investigation (16) dilution of CS by water at half or once its volume was found to reduce ammonia losses by about 25 and 50%, repectively. Further it is also concluded from a laboratory study, that application of the liquid phase of a separated CS diminishes ammonia emission by about 50%. This reduction of ammonia emission is due to an improvement of the flow and infiltration properties of the slurry.

It was shown by a field study that yields and N recoveries of winter wheat are substantially higher, if CS has been stabilized with sulphuric acid. Leaching of sulphur into surface and ground waters, however, may lead to environmental problems (H S).

Although a tow hose spreader and a slurry separator are relatively expensive, the combination of the two systems is considered to be a very effective and economic way to increase the manurial value of cattle slurry.

REFERENCES

(1) SPRENGEL, C. (1839): Lehre vom Dünger, Leipzig

(2) HECK, A.F. (1931).Conservation and availability of the nitrogen in farm manure. Soil Science 31, 335-363

(3) DÖHLER, H., (1988): Bei der Gülleausbringung kann viel Stickstoff verloren gehen. Landwirtschaftsblatt Weser-Ems, 33, 14-16

(4) SMITH, K.A., UNWIN, R.J. and WILLIAMS, J.H. (1985): Experiments on the fertilizer value of animal waste slurries. In: Long term Effects of Sewage Sludge and Farm Slurries Applications (ed.: Williams, Guidi und L'Hermite) London: Elsevier Applied Science, 124-135

(5) NIHLGARD, B.,(1985): The ammonium hypothesis-an additional explanation to the forest dieback in Europe. Ambio 14,1, 2-8

(6) KAUPENJOHANN, M., DÖHLER, H. and BAUER, M., (1989): Effects of N-immissions on nutrient status and vitality of Pinus sylvestris near a hen house. Plant and Soil, 113, 279-282

(7) ROELOFS, J. G. M., KEMPERS, A. J., HONDIJK, A. L. F. M. and JANSEN, J.,(1984): The effect of air-borne ammonium sulfate on Pinus nigra var. maritima in the Netherlands. Plant and Soil, 84, 45-56

(8) BEAUCHAMP, E.G., KIDD, G.E. and THURTELL, G.W., (1978): Ammonia volatilization from sewage sludge applied in the field. J. Environ. Qual., 7, 141-146

(9) RYDEN, J.C. and MCNEILL, E. (1984): Application of the micrometeorological mass balance method to the determination of ammonia loss from a grazed sward. J. Sci. Food Agric., 35, 1297-1310

(10) RANK, M., HUBER, J. and AMBERGER, A. (1988): Model trials on the volatilization of ammonia following slurry application under controlled climate and field conditions. Proceedings of 4th Int. Symposium of CIEC, Braunschweig, Mai 1987,2, 315-320

(11) HOFF, J.D., NELSON, D.W. and SUTTON, A.L., (1981): Ammonia volatilization from liquid swine manure applied to cropland. J. Environ. Qual. 10, 90-95

(12) HOLZER, U., DÖHLER, H. and ALDAG, R., (1988): Ammoniakverluste bei der Rindergülleausbringung im Modellversuch. VDLUFA- Schriftenreihe, 23, Kongreßband 1987, 265-278

(13) CHRISTENSEN, B.T., (1988): Ammonia loss from surface applied animal slurry under sustained drying conditions. In: Volatile emissions from livestock farming and sewage operations. Elsevier Applied Science. I Nielsen (ed.) 92-102

(14) DÖHLER, H. and WIECHMANN, M., (1988): Ammonia volatilization from liquid manure after application in the field. Proceedings of 4th Int. Symposium of CIEC, Braunschweig, Mai 1987, 2, 305-313

(15) THOMPSON, R.B., RYDEN, J.C. and LOCKYER, D.R. (1987): Fate of nitrogen in cattle slurry following surface application or injection to grassland. J. of Soil Sci., 38, 689-700

(16) BEUDERT,B., DÖHLER, H. und ALDAG, R.,(1989): Ammoniakverluste aus mit Wasser verdünnter Rindergülle im Modellversuch. VDLUFA-Schriftenreihe, 28 Kongreßband 1988, Teil II, 1355-1364

EFFECT OF DRY MATTER CONTENT ON AMMONIA LOSS
FROM SURFACE APPLIED CATTLE SLURRY

S.G. SOMMER and B.T. CHRISTENSEN

Askov Experimental Station
Vejenvej 55, DK-6600 Vejen
Denmark

Summary
       The ammonia loss from surface applied cattle slurry was measured
with a wind tunnel system, whereby parameters affecting the volatilization loss can be examined under controlled conditions. The effect of
dry matter content was determined using a slurry adjusted to different
contents of dry matter. The slurry was prepared by mixing the fibrous
and liquid fractions of a mechanically separated slurry.
       Slurry was applied to a 5 cm heigh grass ley at a rate of 3 $1/m^2$.
The content of dry matter was varied from 2.8% to 15.6% with a resulting variation in total-N of 3.1 to 4.9 g N/l. Average pH was 7.7
(C.V. 2%) and ammonia content 2.7 g NH_4-N/l (C.V. 12%). Temperature
varied from -0.2 to 15.7°C and the wind speed (in the steel duct)
ranged from 2.8 to 3.7 m/s.
       The cumulated loss of ammonia over a period of 6 days ranged from
26 to 100% of the applied ammonium. The loss of ammonia was related
linearly to the content of dry matter. After 6 hours of exposure the
relation was affected by climatic conditions, producing a low coefficient of correlation (R^2= 0.46). After 6 days of exposure, interactions was reduced and the squared coefficient of correlation was 0.81.

1. INTRODUCTION
       The intensification of farming has led to accumulation of animal manure
on specialised livestock farms. For this reason and due to the low costs of
fertilizer-N there has been a tendency to neglect losses of nitrogen after
application of manure. Volatilization of ammonia may account for more than
50% of the ammonium in surface applied slurry (6; 12). The deposition of
ammonia may cause detrimental effects on natural ecosystems (9).
       In recent publications (3; 7) ammonia volatilization has been shown to
be higher from cattle than from pig slurry. This was thought to be due to
a lower dry matter content of pig slurry. Wind tunnel experiments with different pig slurries confirmed this (10). However, variations in clima, soil
conditions and pH of the slurry resulted in a weak relationship between the
ammonia loss and the content of dry matter.
       In this study, the ammonia loss was related to the content of dry mat-
ter using a single type of cattle slurry, adjusted to different contents of
dry matter. The slurry was prepared mixing the fibrous and liquid fractions
of a mechanically separated cattle slurry.

2. METHODS
Wind tunnel system
       The wind tunnel system consists of four wind tunnel units, an environ-
mental data monitoring unit, and a gas trapping unit.
       The design of the wind tunnel unit follows essentially that described

by Lockyer (5). The unit consists of an U-shaped tunnel made from transparent polycarbonate and covering the experimental plot (2.0 x 0.5 m). The highest point of the transparent tunnel is 0.45 m above ground. Connected to the tunnel is a circular steel duct housing an electrical powered fan. The fan draws an air stream across the experimental plot and can be controlled to produce wind speeds up to 5 m s^{-1} in the steel duct. The steel duct is 2 m long and has a diameter of 0.4 m.

Temperature within the steel duct is measured with a thermocouple sensor and wind speed with a vane anemometer head. Signals from the anemometer head are led to a display and therefrom to a data-logger. Thereby the anemometers may be used for adjusting tunnel wind speed to pre-selected levels by regulating fan speed. The signals from anemometers and thermo-elements are scanned every 1 minute and stored as 1 hour means on the data-logger.

The air stream is sampled near the outlet of the steel duct by drawing air continuously at a rate of 5 l min^{-1} through a 100 ml absorption flask fitted with a sintered glass distribution tube. The flask contains 50 ml of 0.005 N H_3PO_4. The air flow is checked by a flow-meter. Ammonia in the air is trapped in the orthophosphoric acid and determined with a Berthelot reaction.

Each gas trapping unit is equipped with four absorption flasks. A programmable magnetic valve control automatically switch each absorption flask on and off at pre-selected intervals. The units are placed as close as possible to the sample point to reduce any sorption of ammonia that might occur in the tubes. During the first two days absorption periods were 6 hours; during the following 4 days absorption periods were 24 hours.

Calculation of ammonia loss.
Each experimental run included one untreated reference plot for measurements of background ammonia concentrations and three slurry treated plots. For a given period, background ammonia concentrations were subtracted from those obtained for treated plots:

$$J(hour) = A * V * 3600 * (NH_3,e - NH_3,r)$$

J(hour) : Ammonia loss, g NH_3-N/hour.
A : Area of steel duct, m^2.
V : Wind speed, m/s.
NH_3,e : NH_3 concentration in air from a treated tunnel, g NH_3-N/m^3.
NH_3,r : NH_3 concentration in air from the reference tunnel, g NH_3-N/m^3.

Thus the values presented are net ammonia losses from treated plots. The results are corrected by the recovery factor 100/74 determined in a previous study (4).

3. SLURRIES
Slurries with different contents of dry matter were made by mixing the fibrous and liquid fractions of a mechanically separated cattle slurry. The resulting content of dry matter ranged from 2.8 to 15.6%. Since organic N in the slurry is related to content of dry matter, total-N varied from 3.1 to 4.9 g N/l (Table 1). The mean ammonium concentration was 2.7 g N/l (C.V. 12%) and pH was 7.7 (C.V. 2%). Ammonium concentration is slightly related to dry matter content, whereas pH was not related to dry matter content.

Table 1. Characteristics of the slurries used in the experiments.

Exp. no.	Exp. period	Dry matter %	N-total g N/l	NH_4^+-N g N/l	pH
1	Aug. 1989	6.9	3.1	1.7	7.4
2	Aug. 1989	4.1	3.3	2.2	7.5
3	Aug. 1989	3.6	3.7	2.6	7.8
4	Oct. 1989	2.8	3.9	2.7	7.7
5	Oct. 1989	8.2	4.2	2.8	7.6
6	Oct. 1989	15.6	4.9	2.9	7.9
7	Oct.-nov.1989	2.8	3.9	2.7	7.7
8	Oct.-nov.1989	8.2	4.2	2.8	7.6
9	Oct.-nov.1989	15.6	4.9	2.9	7.9
10	Nov. 1989	5.2	4.4	3.0	7.7
11	Nov. 1989	6.0	4.3	2.9	7.7
12	Nov. 1989	10.0	4.6	2.9	7.6
13	Nov. 1989	5.2	4.4	3.0	7.7
14	Nov. 1989	6.0	4.3	2.9	7.7
15	Nov. 1989	10.0	4.6	2.9	7.6

Fifteen experimental runs was carried out. The slurry was surface applied at a rate of 3 l/m^2 on a 5 cm high grass ley, grown on a sandy loam with 10 % clay and 2.7 % organic matter.

4. AMMONIA LOSS

The experiments was carried out in the period from august to november 1989, where the recorded air temperature was between -0.2 and 15.7°C. The wind speed in steel duct was in the interval from 2.8 to 3.7 m/s (Table 2). Atmospheric ammonia concentrations measured in the untreated reference tunnel was 2.9-36.2 ug NH_3-N m^{-3}, and was similar to the level often found in areas with livestock (1; 2).

More than half of the ammonia loss occured within one or two days. Figure 1 indicates that the loss of ammonia from surface applied slurry follows an exponential curve.

The cumulated loss of applied ammonium during 6 days was 26% and 100% for slurries with a content of dry matter of 2.8% and 15.6%, respectively (Table 2). These losses fall within the range of ammonia losses reported elsewhere (12; 8).

The ammonia loss from slurries having 3.6, 4.1 and 6.9 % of dry matter is illustrated in figure 1. After 6 hours the ammonia loss differed significantly between slurries with different contents of dry matter. The differences in the ammonia losses after 6 days were higher than after 6 hours, indicating that the content of dry matter also affected the ammonia loss during the period of day 1 to day 6.

The typical changes in the flux of ammonia from surface applied slurry may be explained by changes in proton activity occuring in the surface of the slurry (11). The results suggest that the loss of ammonia by volatilization can be divided in two phases. An initial phase, including the first one or two days after application of slurry, where climatic conditions largely determine the loss rate, and a second phase where other factors control the loss rate.

Table 2. Average wind speeds, air temperatures and cumulated ammonia losses from cattle slurries with different dry matter contents
(standard deviations in brackets).

Exp. no.	Windspeed* m/s	Air temp. °C	NH$_3$ loss after 6 days** %
1	3.2 (0.05)	15.7 (1.8)	57
2	2.8 (0.04)	15.7 (1.8)	42
3	2.8 (0.03)	15.7 (1.8)	30
4	3.7 (0.08)	10.9 (1.4)	26
5	3.7 (0.06)	10.9 (1.5)	38
6	3.4 (0.09)	10.9 (1.5)	81
7	3.6 (0.08)	7.6 (1.9)	41
8	3.6 (0.05)	7.5 (1.9)	65
9	3.3 (0.09)	7.5 (1.9)	100
10	3.3 (0.05)	1.5 (2.6)	38
11	3.1 (0.07)	1.5 (2.7)	45
12	3.2 (0.04)	1.5 (2.6)	86
13	3.3 (0.04)	-0.2 (2.3)	41
14	3.1 (0.05)	-0.2 (2.2)	46
15	3.2 (0.03)	-0.2 (2.2)	72

* Measured in the steel duct.
** Cumulated loss of ammonia, p.c. of applied ammonia.

Figure 1. Cumulated ammonia loss from surface applied slurry,
p.c. of applied ammonium.

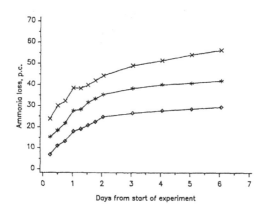

Ammonia loss, p.c.

Days from start of experiment

✕—✕—✕ 6.9 ✕ DM   ✱—✱—✱ 4.1 ✕ DM   ◇—◇—◇ 3.6 ✕ DM

Figure 2. Cumulated ammonia loss from surface applied slurry after 6 hours of exposure, p.c. of applied ammonium.

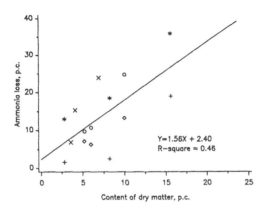

Y=1.56X + 2.40
R-square ≈ 0.46

——— Regression line	x x x Expt. nr. 1–3	+ + + Expt. nr. 4–6
* * * Expt. nr. 7–9	o o o Expt. nr. 10–12	◇ ◇ ◇ Expt. nr. 13–15

    This explains why the squared coefficient of correlation after 6 hours is as low as 0.46 for the relation between the cumulated loss of ammonia and the content of dry matter (Figure 2). After 6 hours of exposure, the ammonia loss measured in experiments 4 to 6 and 7 to 9 was lower and higher respectively, than could be expected from looking at the dry matter content of the slurry. Probably the interaction of environmental conditions is reduced during the second phase of exposure as the cumulated loss of ammonia after 6 days of exposure is significantly ($R^2 ≈ 0.81$) correlated to the dry matter content of the slurries (Figure 3).
    Losses of ammonia from surface applied slurry is related linearly to content of dry matter. The coefficient of correlation for the relationship approaches one with increased exposure time, when all other parameters characterising the slurries is constant. This suggests that there is a time dependent increasing importance of content of dry matter on ammonia loss from slurry left on the soil surface.

Figure 3. Cumulated ammonia loss from surface applied slurry after 6 days of exposure, p.c. of applied ammonium.

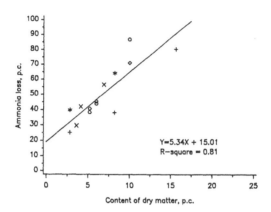

Content of dry matter, p.c.

———— Regression line	× × × Expt. nr. 1–3	+ + + Expt. nr. 4–6
* * * Expt. nr. 7–9	o o o Expt. nr. 10–12	◇ ◇ ◇ Expt. nr. 13–15

Acknowledgement
Financial support for this study was obtained from The National Agency of Environmental Protection as part of the N, P and Organic Matter Research Program 1985-1990.

5. REFERENCES

(1) ASMAN, W.A.H., PINKSTERBOER E.F., MAAS, H.F.M., ERISMAN, J.W., YPE-LAAN, A.W., SLANINA, J. and HORST, T.W. (1989). Gradients of the ammonia concentration in a nature reserve: Model results and measurements. Atm. Environ. 23,2259-2265.

(2) ALLEN, A.G., HARRISON, R.M. and WAKE, M.W. (1988). A meso-scale study of atmospheric ammonia and ammonium. Atm. Environ. 22, 1347-1353.

(3) CHRISTENSEN, B.T. (1988). Ammonia loss from surface-applied animal slurry under sustained drying conditions in autumn. In Nielsen V.C. et al. (eds.): Volatile Emissions from Livestock Farming and Sewage O-perations. Elsevier Applied Science. London & New York, 92-102.

(4) CHRISTENSEN, B.T. and SOMMER, S.G. (1989). Fordampning af ammoniak fra udbragt gødning. Metode og ammoniaktab fra urea og urea-ammonium-nit-rat. (Volatilization of ammonia from fertilizers and manure. Methodology and loss of ammonia from urea and urea-ammonium-nitrate, in Danish with English summary). Tidskr. Planteavl 93, 177-190.

(5) LOCKYER, D.R. (1984). A system for the measurement in the field of losses of ammonia through volatilization. J. Sci. Food Agric. 35, 837-848.

(6) LOCKYER, D.R., PAIN, B.F. and KLARENBEEK, J.V. (1989). Ammonia emissions from cattle, pig and poultry wastes applied to pasture. Environ. Pollut. 56, 19-30.

(7) PAIN, B.F., REES, Y.J. and LOCKYER, D.R. (1988). Odour and ammonia emission following the application of pig or cattle slurry to land. In Nielsen V.C. et al. (eds.): Volatile Emissions from Livestock Farming and Sewage Operations.Elsevier Applied Science. London & New York, 2-11.

(8) PAIN, B.F., THOMPSON R.B., REES Y.J. and SKINNER, J.H. (1990). Reducing gaseous losses of nitrogen from cattle slurry applied to grassland by the use of additives. J.Sci. Food Agric. 50, 141-153.

(9) ROELOFS, J.G.M. (1986). The effect of airborne sulphur and nitrogen deposition on aquatic and terrestrial heathland vegetation. Experientia 42, 372-377.

(10) SOMMER, S.G. and CHRISTENSEN, B.T. (1989). Fordampning af ammoniak fra svinegylle udbragt på jordoverfladen. (Volatilization of ammonia from surface-applied pig slurry, in danish with English summary). Tidskr. Planteavl 94, 345-359.

(11) SOMMER, S.G., OLESEN, J.E. and CHRISTENSEN, B.T. (1990). Effect of temperature, wind speed and humidity on ammonia loss from surface applied cattle slurry. In preparation.

(12) THOMPSEN, R.B., RYDEN, J.C. and LOCKYER, D.R. (1987). Fate of nitrogen in cattle slurry following surface application or injection to grassland. J. Soil Sci. 38, 689-700.

FIELD INVESTIGATION OF METHODS TO MEASURE
AMMONIA VOLATILIZATION

M. FERM[1], J.K. SCHJØRRING[2], S.G. SOMMER[3], and S.B. NIELSEN[2]

[1]Swedish Environmental Research Institute,
P.O. Box 47086, S-402 58 Gothenburg.

[2]Royal Veterinary and Agricultural University,
Dept. of Soil, Water and Plant Nutrition, Thorvaldsensvej 40,
DK-1871 Frederiksberg C, Denmark

[3]Askov Experimental Station, Vejenvej 55, DK-6600 Vejen, Denmark

Summary
A new type of passive sampler for determination of ammonia fluxes from land surfaces or from manure stores have been developed. The purpose of this project was to compare it to a conventional mass balance technique on a small plot with a known ammonia emission. The known emission was achieved by placing 140 beakers containing an ammonium sulphate solution on a circular plot. Sodium bicarbonate was then added to all the beakers and ammonia started to volatilize. The experiment was stopped after 9 hours and sulphuric acid was added to the beakers. The amount of ammonia lost from the beakers was compared to that determined by passive flux samplers on the circumference of the plot and to that determined with conventional ammonia samplers (acid traps) with wind speed equipment, in the middle of the plot. Two experiments have been carried out so far.

1. INTRODUCTION

There is a great need for data on ammonia volatilization from manure stores and from manure applied to the surface of agricultural land. Furthermore, very few results are available on the emission of ammonia from the foliage of agricultural crops.

The ammonia volatilization process is strongly influenced by climatic parameters like air temperature, wind speed, and air humidity. The frequently used chamber technique alters the micro- climatic conditions and may give incorrect estimates of the ammonia emission (1,2). Instead, micrometeorological techniques have to be used because they do not disturb the environmental conditions which influence ammonia volatilization (3,4).

Micrometeorological measurements based on energy balance or aerodynamic (gradient diffusion) methods require extensive and uniform land areas. Such requirements are not necessary in micro- meteorological measurements based on mass balance (3). In the mass balance approach, the flux of the gas under consideration is determined from the differences in the amount of gas driven by the wind across the windward and leeward boundaries of the experimental area. No special form of the wind speed profile or correction for thermal instability is required. Micro- meteorological mass balance methods are thus suitable for measuring ammonia emissions from smaller plots. However, in their conventional form (see e.g. 5) they are very labour demanding and still have the disadvantages associated with requirements of instrumentation for measurements of the NH_3-concentration in the air, wind speed and wind direction.

In order to simplify micro-meteorological mass-balance measurements a simple technique based on passive ammonia samplers have recently been developed (6). This technique has several potential advantages:

- easy to operate
- low labour requirements
- ammonia emission can be integrated over long time periods
- wind speed and wind direction measurements are not required
- electricity is not needed

The passive samplers collect ammonia in an amount that is proportional to the product of the ammonia concentration along the sampler and the wind speed composant along the sampler (6). The collected amount of ammonia is thus proportional to the integrated horizontal ammonia flux through a fixed vertical surface. By adding the ammonia flux that leaves the experimental area through four directions perpendicular to each other at several heights (Fig.3) and subtracting the corresponding fluxes that enters the field from the surroundings, the ammonia volatilization from the plot can be estimated.

In the present investigations, the technique with passive ammonia samplers was tested against a conventional micro-meteorological mass-balance method on an artificial field with a known ammonia emission rate.

2. EXPERIMENTAL

Experimental area

The experiments were carried out on a not fertilized ploughed field just outside Copenhagen, Denmark at the experimental farm belonging to The Royal Veterinary and Agricultural University. The experimental area was a circle with an area of 707 m^2 (radius 15 m). The distance from the experimental area to hedges, houses etc. was more than 100 m.

Ammonia source

The ammonia source consisted of 140 flat beakers (diameter 30 cm, height 4 cm), distributed evenly on the area and containing 0.3 litre 1.2 M NH_4^+ solution (experiment 1) or 0.5 litre 1.0 M NH_4^+ solution (experiment 2). Besides NH_4^+ , the solution contained 0.6 M $NaHCO_3$. The NH_4^+ /HCO_3^- solution was made at the start of the experiments by adding $NaHCO_3$ solution to a $(NH_4)_2SO_4$ solution already in the beakers. The pH of the NH_4^+ /HCO_3^- solution was initially 8.5 and did not drop below 8 during the experiments. The amount of NH_3 volatilized was estimated on basis of the difference in NH_4^+ content in 50 ml samples taken from the beakers before and after an experiment. Immediately after the samples were taken, they were acidified by addition of 50 ml of 0.6 M H_2SO_4 in order to arrest further losses of NH_3. The NH_4^+ concentration in the samples was determined by distillation followed by titration. The amount of water evaporated from the beakers during the experiments was estimated from measurements of the volume of the NH_4^+ /HCO_3^- solution before and after the experiments.

Mass balance method with passive flux samplers

The passive flux samplers consisted of two pairs of glass tubes connected in series (see fig 1). The inside was coated with oxalic acid by means of a 3% solution in methanol. The tubes were 100 mm long having an internal diameter of 7 mm and an external of 10 mm. They were connected in series with each other and a probe by means of a short piece of silicon tubing. One unit samples NH_3 when the air blows from the probe end and the other glass tube samples NH_3 when the wind comes through the open end. The purpose of the probe is to decrease the wind speed inside the tubes in order to achieve a low friction resistance and a high NH_3 collection efficiency. The wind speed inside the tubes will approximately be proportional to the wind speed outside the probe multiplied with cosine for the angle between the wind direction and the axes of the tubes (see Fig 1). The probe consisted of a stainless steal foil with a circular hole in the centre. The thickness of the foil was 0.5 mm. A thickness of 0.05 mm would probably have been better (see 6).The hole in the foil was made using a laser, a procedure which gave a high precision compared to the previously used punch method (7). Tubes and probes were manufactured by Mikrolab Aarhus, A/S Axel Kiers Vej 34, DK-8270 Hoejbjerg, Denmark. When the tubes are not being exposed they are capped in both ends by plastic caps. Clean caps are also used during the leaching of the tubes. 3 ml deionized water is than added and the tubes are shaked very carefully. NH_4^+ analysis of the leachate is made using flow injection analyses (FIA).

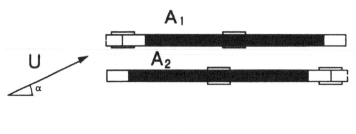

$$\frac{(A_1 + A_2)}{2 \cdot \pi \cdot r^2 \cdot 0.7} \approx \int_{t_1}^{t_2} U \cdot \cos \cdot \alpha \cdot [NH_3] \cdot dt$$

Fig 1 Passive flux samplers for ammonia

The vertical flux is calculated from the average amount of ammonia in the tubes that faces the same direction, divided by the area of the hole and divided by a correction factor (0.7). The correction factor has been determined experimentally in a wind tunnel. The factor is smaller than unity, because the probe creates a turbulence that decreases the wind speed. The blank of an unexposed tube is about 0.06 µg N corresponding to a flux of 0.1 g N m^{-2}. The analytical detection limit is about half of the blank. The ammonia loss from the experimental plot (X) is calculated by multiplying all outgoing fluxes ($\emptyset_F$)with the width of the field (W) and the height interval that the sampler represents (Δh) and subtracting the corresponding incoming amounts for all heights (h) and masts (m).

$$X = \sum_{m=1,h=0}^{m=4,h=H} \emptyset_F \cdot W \cdot \Delta h \ - \ \sum_{m=1,h=0}^{m=4,h=H} \emptyset_S \cdot W \cdot \Delta h \qquad (1)$$

In experiment 1 (September 28th, Table 1) the four masts with the passive NH$_3$ were placed on a square with a side length of 21 m. The square was inscribed in the circular experimental area. In experiment 2 (November 16th, Table 2) the four masts were placed on the circumference of the experimental circle (Fig. 3). This is a more suitable configuration. Four heights, namely 75, 150, 225, and 300 cm above the soil surface were used at each of the 4 masts. Since the four masts are placed in the middle of the sides and the enclosed plot is circular, the masts will receive a flux that is slightly higher than the average flux through the side.

The wind travels an average distance over a circular equal to the area divided by the width ($\pi r^2/2r = \pi r/2$). Two pairs of flux samplers recieves air that has passed a distance of $2 \cdot r \cdot \cos \alpha + 2 \cdot r \cdot \sin \alpha$ (see Fig 2). The flux is not only proportional to the NH$_3$ concentration (here assumed to be proportional to the distance travelled above the source) and the wind speed but also to cosine for the angle between the wind direction and the axes of the samplers. In this case the product will be $2 \cdot r \cdot \cos^2 \alpha$ and $2 \cdot r \cdot \sin \alpha \cdot \cos(90 - \alpha)$ respectively. When these two fluxes are added in eq. 1 the sum will be $2 \cdot r$. The masts placed on the middle of the sides of a square surrounding a circular plot therefore recieves a flux which is higher than the integrated flux along the sides ($\pi r/2$). To compensate for this "over representativeness" the lost NH$_3$ amount must be divided by $4r^2$ instead of πr^2 in order to get the amount lost per unit area.

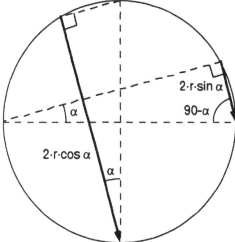

Fig 2. Graphic presentation of the representativeness of the four masts surrounding the circular plot. The figure is explained in the text.

<u>Conventional mass balance method</u>

The leeward mast, placed centrally in the experimental area, was equipped with an ammonia trap in 25, 50, 75, 110, 150, 225, and 375 cm above the soil surface (Figure 3). A mast with an air temperature and air humidity sensor (Rotronic MP100TST- 010 Meteorol. probe) in 100, 150, 200, and 300 cm height above soil surface and an anemometer (Vector Instruments A 101M) in 25, 50, 75, 110, 150, 225, and 375 cm height above soil surface was placed in 1 m distance from the central mast. The signals from the air temperature probes and from the anemometers was recorded by a data logger (Cambell CR10/WP). Two windward masts with an ammonia trap in 100, 200, and 300 cm height above the soil surface were placed outside of the experimental area (Figure 3).

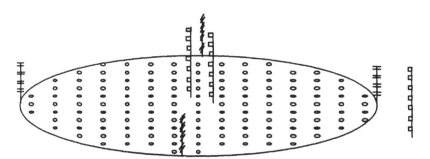

Figure 3 The experimental plot with 140 beakers containing NH_3. The four masts used for the passive flux samplers are located on the periferi of the circular plot. The mast with the conventional NH_3 samplers and the wind sensors, are located in the middle.

The NH_3 trap consisted of a 250 ml Dreschel test tube with a bottle head supporting a gas dispersion tube (pore size 1). The traps contained 60 ml 0.25 M H_2SO_4. The air flow through the trap was 8 l min^{-1} provided by a Neuberger diaphragm pump (model N 79KN.18). The flow was checked in the field with a Brooks sho- rate purgemeter. The traps were changed every 3rd hour. The concentration of ammonium in the traps was measured colorimetrically with an autoanalyzer. The method was based on the phenylhypo-chlorite-ammonium reaction of Berthelot, increased in rate and intensity by Na-prusside.

The ammonia loss from the experimental area was calculated by using equations similar to those presented by Ryden and McNeill (5). In experiment 2, the relationships between the logarithm of height and the leeward concentrations of NH_3 were not linear. The integration with respect to height was in this case carried out by dividing the concentration profile in two parts that each could be linearized.

3. RESULTS AND DISCUSSION

The amount of NH_3 volatilized in a period of 9 hours in experiment 1 was 374±26 mg NH_3-N m^{-2} (Table 4). In experiment 2, only 34 mg NH_3- N m^{-2} was lost during the 9 hours experimental period (Table 4). The difference in NH_3-flux between the two experiments was likely due to a higher air temperature in experiment 1 than in experiment 2 (Table 3), as ammonia loss from an aqueous solution is exponentially related to temperature (8). In addition, a low wind speed in experiment 2 may have reduced the NH_3-emission.

The ammonia concentration at each sampling position above the experimental surface increased considerably during the course of experiment 1. The same was the case for the background NH_3-concentration which rose from 10 μg NH_3-N m^{-3} in the first 3-hour sampling interval (8 to 11 a.m.) to 35 μg NH_3-N m^{-3} in the last 3-hour sampling interval (2 to 5 p.m.). The reason for the increase in background NH_3 is unknown. The height z_p in which the aerial NH_3 concentration above the surface of the experimental area was estimated to be equal to the background NH_3 concentration was 3.7, 2.9 and 0.7 m for sampling intervals 1, 2 and 3, respectively. The decline in z_p probably reflects the creation of more stable atmospheric conditions near the end of the experimental period due to a change in the air temperature gradient in this period, leading to increasing air temperature with height in the last part of the period (data not shown). In addition, the high background NH_3-levels by the end of the experiment may have overridden the effects of the NH_3 source in heights above z_p.

In experiment 2 the ammonia concentration in each height above the experimental surface varied very little during the experimental period (standard errors for the mean concentration in each height during the experimental period were less than 7%). The background NH_3 concentration was in all cases less than 1 μg NH_3-N m^{-3}. Average z_p was 1.7±0.1 m .

The total NH_3 emission determined by the micro-meteorological mass balance technique with conventional NH_3 traps was 356 mg NH_3-N m^{-2} (9 hours)$^{-1}$ in experiment 1 and 19 mg NH_3-N m^{-2} (9 hours)$^{-1}$ in experiment 2 (Table 4). Distributed on the 3 three-hour sampling intervals during the experimental period (8- 11 a.m., 11 a.m.-2 p.m., and 2-5 p.m.) the losses were 27, 44 and 29% of the total in experiment 1, and 31, 43 and 27% of the total in experiment 2, respectively. The NH_3 flux determined by the conventional micro-meteorological mass-balance technique was in experiment 1 very close to the amount of NH_3-N volatilized from the beakers with NH_4^+/HCO_3^-- solution (Table 2). In experiment 2 with the very low NH_3 flux density, the NH_3 flux determined by the micro-meteorological measurements was only 55% of that determined on basis of a NH_4^+ mass-balance for the beakers with the NH_4^+/HCO_3^- solution (Table 4).

The NH_3 emission measured by the flux samplers was in the first experiment 28% lower than that determined from the analysis of the beakers (see Table 4). There are two possible explanations to this.

1) The foil in the probe was much too thick. This fact will lower the flow inside the glass tubes when the air has to make a bend in the hole. This was the case in the first experiment when the fluxes from the field was approximately equal at two masts (45° angle, see Table 1, East & South). However in the second experiment most NH_3 was transported through the south side (Table 2) i.e. perpendicular to the side.

2) In the first experiment the four masts were placed on the middle of the sides of a square enclosed in the circle. There were consequently beakers outside the field, which contributed to the background. A high background from the surroundings was also observed in this experiment (Table 1 (s)). If some beakers were too close to a flux sampler they could have affected the result in a none representative way. If one assumes that the fluxes from the surroundings at 0.75m height are too high and uses a a flux equal to the average flux of the other heights, a background flux of 57 g N will be obtained. This background corresponds to a total emission of 322 mg N m^{-2}.

Table 1 Horizontal fluxes from the field (F) and the surroundings (S) as a function of mast and height during the first experiment. Fluxes have also been multiplied with the vertical areas that they represent (Total). The emission from the field was $(199-80)/21^2 = 0.27$ g N m^{-2}.

height	Fluxes in g N m^{-2} North		East		South		West		Total (g N)	
	F	S	F	S	F	S	F	S	F	S
300	0.00	0.22	0.44	0.00	0.55	0.25	0.11	0.38	17.2	13.3
225	0.08	0.16	0.71	0.19	0.57	0.19	0.03	0.25	21.9	12.5
150	0.00	0.38	1.28	0.25	0.79	0.19	0.00	0.27	32.6	17.2
75	0.08	0.55	3.00	0.27	2.13	0.14	0.19	0.60	128	36.7
Total	3.2	24.9	109	13.3	80.3	13.1	6.7	28.4	199	80

Table 2 Horizontal fluxes from the field (F) and the surroundings (S) as a function of mast and height during the second experiment. Fluxes have also been multiplied with the vertical areas that they represent (Total). The emission from the field was $(46-17)/30^2$ =0.032 g N m^{-2}.

height	Fluxes in g N m^{-2} North		East		South		West		Total (g N)	
	F	S	F	S	F	S	F	S	F	S
300	0.00	0.05	0.05	0.03	0.19	0.05	0.05	0.03	6.7	3.7
225	0.03	0.08	0.11	0.05	0.14	0.05	0.05	0.05	7.3	5.5
150	0.00	0.05	0.05	0.00	0.22	0.03	0.14	0.05	9.2	3.0
75	0.03	0.05	0.00	0.05	0.41	0.03	0.25	0.00	23.0	4.6
Total	1.5	6.1	4.9	3.7	26.1	4.0	13.7	3.0	46	17

Climatic conditions

Mean wind speed, air temperature, and air humidity in various heights above the soil surface are given in Table 3. It is noted that both the air temperature and the wind speed was much higher in experiment 1 than in experiment 2.

Table 3 Wind speed, air temperature and humidity during the two experiments.

Parameter	Height cm	Experiment 1 28/9 '89	Experiment 2 16/11 '89
	25	5.0	1.7
	50	4.6	1.3
Wind speed	75	4.4	1.2
m s^{-1}	110	4.1	1.1
	150	3.8	1.0
	225	3.5	1.0
	375	2.9	0.8
	100	14.5	4.3
Air temperature	150	14.9	4.0
°C	200	14.4	4.1
	300	14.6	3.2
	100	65	70
Air humidity	150	65	72
% r.h.	200	64	71
	300	64	71
Wind direction		SW-NW	NW-NE

Table 4. Comparison of NH_3 fluxes obtained from measured NH_4^+ loss in the 140 beakers, a conventional masss balance technique and the new type of passive flux samplers. The figures are given in mg N m^{-2} during 9 hours.

Date	Beakers	Conventional mass balance	Passive flux samplers
28/9 '89	374±26	356	270
16/11 '89	34±13	19	32

Acknowledgements.

The work was supported by funds from the NPO-programme under the Danish Environmental Protection Agency.

REFERENCES
(1) Ferm, M. 1983. Ammonia volatilization from arable land - An evaluation of the chamber technique. - In Observation and Measurement of Atmospheric Contaminants. WMO Special Environmental Report. 16, 145-172.
(2) Ferguson,R.B., McInnes, K.J., Kissel, D.E. & Kanemasu, E.T. 1988. A comparison of methods of estimating ammonia volatilization in the field. - Fert. Res. 15, 55-69.
(3) Denmead, O.T. 1983. Micrometeorological methods for measuring gaseous losses of nitrogen in the field. In Gaseous Loss of Nitrogen from Plant-Soil systems (J. R. Freney & J. R. Simpson, eds.), pp. 133-157. - Development in Plant and Soil Sciences. 9, Martinus Nijhoff/Dr. W. Junk Publishers, The Hague/Boston/Lancaster.
(4) Black, A.S., Sherlock, R.R., Cameron, K.C., Smith, N.P., and Goh, K.M. 1985. Comparison of three field methods for measuring ammonia volatilization from urea granules broadcasted on to pasture. - J. Soil Sci. 36, 271-280.

(5) Ryden, J.C. & McNeill, J.E. 1984. Application of the micrometeorological mass balance method to the determination of ammonia loss from a grazed sward. - J. Sci. Food Agric. 35, 1297-1310.

(6) Ferm M. 1986. Concentration Measurements and Equilibrium Studies of Ammonium, Nitrate and Sulphur Species in Air and Precipitation. - Ph.D. Thesis, 77 pp., Department of Inorganic Chemistry, Chalmers Tekniska Högskola, Göteborg, Sweden. ISBN 91-7900-006-1.

(7) Ferm, M. & Christensen, B.T. 1987. Determination of NH$_3$ volatilization from surface-applied cattle slurry using passive flux samplers. In Ammonia and Acidification (W. A. H. Asman, and S. M. A. Diederen, eds.), pp. 28-41. National Institute of Public Health and Environmental Hygiene, Bilthoven, The Netherlands.

(8) Vlek, P.L.G. & Stumpe, J.M. 1978. Effects of solution chemistry and environmental conditions on ammonia volatilization losses from aqueous systems. - Soil Sci. Soc. Am. J. 42, 416-421.

MODELLING   AMMONIA   EMISSIONS   FROM   ARABLE   LAND

W.J.CHARDON, J. VAN DER MOLEN* and H.G. VAN FAASSEN

Institute for Soil Fertility
P.O. Box 30003
NL 9750 RA Haren
Netherlands

Summary
   Ammonia, volatilized from animal manure after land-spreading, is
one of the major sources of acid deposition in the Netherlands. A
model is presented which describes the transfer of ammonia from arable
land to the atmosphere after surface application or incorporation of
animal manure. The model can be used to study the interaction of the
chemical, physical and environmental factors influencing volatilization losses and their combined influence on NH_3 volatilization under
field conditions.
The model employs the following flux equation
   $R = k.(C_s - C_a)$
where k is a transfer coefficient, C_s is the surface $NH_{3(g)}$ concentration and C_a is the atmospheric $NH_{3(g)}$ background concentration. The rate of volatilization R can be calculated at any moment
after application, provided k, C_s and C_a are known. The model therefore basically consists of modules which yield these variables.

1. INTRODUCTION

   Annual ammonia (NH_3) emissions from liquid animal manures in the
Netherlands are estimated to be 2.5×10^5 Mg. A substantial part of the NH_3 is
deposited on nearby sites, and contributes to soil acidification upon transformation into HNO_3 through nitrification. It is estimated that volatilization of NH_3 after land-application of cattle slurry constitutes 30 % of the
total emission of NH_3.
   Results of field experiments on NH_3 volatilization after application of
cattle slurry to arable land have been reported by several authors (1,2,3).
As the interactions involved in the NH_3-volatilization process are very complex, modelling the process has become a prerequisite for understanding the
dynamics of the process, and for interpretation of the results of experiments. As far as we know there are no models available describing the process of NH_3 volatilization under field conditions from arable land after application of cattle slurry.
   In this paper, we briefly describe a transfer model for NH_3 volatilization from arable land after surface application or incorporation of cattle
slurry; the model is described more extensively elsewhere (4). The model can
be used to study the interaction of the chemical, physical and environmental
factors influencing volatilization losses and their combined influence on
NH_3 volatilization under field conditions; it can serve as the base for a
predictive NH_3 volatilization model. In an appendix special attention is
given to factors that determine the pH of a soil/manure mixture.

* Present affiliation: Tauw Infra Consult BV, PO Box 479
                        NL 7400 AL Deventer, Netherlands

2. MODEL DESCRIPTION, GENERAL THEORY

Cattle slurry is a mixture of urine and faeces excreted by cattle, with a variable amount of cleaning water added during housing. About 50 % of the nitrogen in the slurry is in the form of urea, which undergoes hydrolysis :
(1) $CO(NH_2)_2 + 3H_2O ---> (NH_4)_2CO_3 + H_2O ---> 2NH_4^+ + HCO_3^- + OH^-$
(2) $NH_4^+ <---> NH_3^0 + H^+$
The NH_3^0 thus formed may volatilize during housing and storage, and during and after land-application. Volatilization occurs when the concentration of NH_3 at the surface exceeds the concentration of NH_3 in the air, otherwise deposition will occur. The rate of volatilization R ($\mu gN.m^{-2}.s^{-1}$) can be calculated as :
(3) $R = k.(C_s - C_a)$
where k ($m.s^{-1}$) is a transfer coefficient, C_s is the concentration of $NH_{3(g)}$ at the soil surface and C_a is the atmospheric $NH_{3(g)}$ background concentration ($\mu gN.m^{-3}$).
The model consists of two main modules : the soil module, where C_s is calculated, and the transfer module, where the value of k is calculated. The value of C_a during the experiment is required as input variable. In the following both modules are briefly described.

3. SOIL MODULE

The distribution of the ammoniacal N, applied with slurry, immediately after application depends on the method of application. In case of surface application slurry is spread on the land after which the slurry infiltrates into the soil, whereas in case of incorporation the slurry is mixed through the upper layer of soil after spreading. For both surface application and incorporation, the initial distribution of the ammoniacal N is assumed to be uniform down to a certain depth. The difference between the two application techniques appears in the depth over which the ammoniacal N is initially distributed. This depth, L_{init} (m), is selected to reflect the principal extent of cattle slurry placement, i.e. in case of surface application the effective distance over which infiltration occurs, and in case of incorporation the thickness of the soil layer through which the slurry is mixed. The assumption of a uniform initial ammoniacal-N distribution with depth down to depth L_{init} implies that in case of surface application infiltration of slurry takes place instantaneously.
For the purpose of modelling, the ammoniacal-N content at the soil surface is assumed to be uniform to some fixed depth, L_1 (m). Thus, throughout a top compartment the temporal variations in ammoniacal-N content resulting from volatilization and chemical, physical, or biological processes in the soil are assumed to be uniform. Below this depth, a second compartment of variable thickness L_2 (m) is assumed from which no volatilization takes place. This compartment acts as a storage reservoir for the amount of ammoniacal N originating from the slurry that is placed below depth L_1. The front of the ammoniacal-N content profile in the soil coincides with the bottom of this compartment at a depth L (m), where a step-change in ammoniacal-N content occurs. Within this compartment the distribution of the ammoniacal N present and the temporal variations in ammoniacal-N content due to chemical, physical or biological processes are also assumed to be uniform with depth.
As the initial distribution of the ammoniacal N is assumed to be uniform down to depth L_{init}, the amounts of ammoniacal N initially stored in the two compartments, $NH_{x1.i}$ and $NH_{x2.i}$ (both in $\mu gN.m^{-2}$), are defined by

the total amount of ammoniacal N applied with the slurry, $NH_{x.app}$ ($\mu gN.m^{-2}$), the depth L_{init} over which this ammoniacal N is placed initially and the thickness L_1 of the top compartment (L_1 is a fixed value for a certain type of application; its magnitude is obtained from calibration of the model). The model therefore requires $NH_{x.app}$, L_{init} and L_1 as input, and calculates the initial value of L_2, $L_{2.i}$, and $NH_{x1.i}$, $NH_{x2.i}$ as follows:

(4) $L_{2.i}$ = $L_{init} - L_1$

(5) $NH_{x1.i}$ = $NH_{x.app}.L_1$ /L_{init}

(6) $NH_{x2.i}$ = $NH_{x.app}.L_{2.i}/L_{init}$

Figure 1 gives a schematic representation of the distribution of the total amount of ammoniacal N originating from the slurry immediately after application.

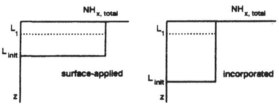

Fig. 1 Schematic representation of the initial distribution of NH_4-N originating from the slurry following surface application or incorporation of the same amounts of slurry

The volumetric flux of water, J_w ($m.s^{-1}$), through the two compartments, which is assumed to be constant with depth, is calculated from the net difference between the evaporation rate, E ($m.s^{-1}$), and the precipitation (=rainfall) rate, P ($m.s^{-1}$), i.e.

(7) $J_w = E - P$

Data on evaporation and rainfall rates throughout the event are required as input for the model.

The flux of ammoniacal N between the two compartments is assumed to take place by convective transport in the liquid phase and diffusive transport in both the liquid and the gas phase.

Convective transport, J_c ($\mu gN.m^{-2}.s^{-1}$), is calculated as :

(8) $J_c = J_w$. ($[NH_3^0]_{aq} + [NH_4^+]_{aq}$)

where $[NH_3^0]_{aq}$ and $[NH_4^+]_{aq}$ are the concentrations ($\mu gN.m^{-3}$) of NH_3^0 and NH_4^+ in solution. Ammoniacal N is transported from the bottom compartment to the top compartment if the net water flux is in the upward direction, E>P, whereas transport of ammoniacal N from the top compartment to the bottom compartment takes place in case of a downward net water flux, E<P. The way $[NH_3^0]_{aq}$ and $[NH_4^+]_{aq}$ are calculated is shown below.

Diffusive transport, J_d ($\mu gN.m^{-2}.s^{-1}$), is calculated as the sum of gaseous diffusion of NH_3 and of aqueous diffusion of NH_4^+ and NH_3^0, which are in turn calculated from the differences in concentration between the two compartments, e.g for NH_4^+ :

(9) $J_d = -D_{aq}$. ($[NH_4^+]_{aq,1} - [NH_4^+]_{aq,2}$) . L_d^{-1}

Subscripts 1 and 2 refer to compartments 1 and 2. The diffusion length L_d is calculated as $L_d = [L_1+L_2]/2$. Effective diffusion coefficients for soil are calculated using the expressions (see Jury et al (5)) :

(10a) $D_{aq} = (\theta_v^{10/3}/\theta_t^2).D_{aq}^{water}$

(10b) D_g = $(\theta_g^{10/3}/\theta_t^2).D_g^{air}$

where D_{aq}^{water} and D_g^{air} are the water-liquid and the air-gas diffusion coefficients ($m^2.s^{-1}$) and θ_v, θ_g and θ_t are the volumetric moisture content, the gas-filled pore fraction and total porosity. Values of $D_{aq}{}^{water}$ and $D_g{}^{air}$ are dependent of temperature, according to a relationships derived from data for water of Yuan-Hui & Gregory (6) :

(11) $D_{aq}{}^{water}$ = 9.8 . 10^{-10} . $1.03^{(T-273)}$
(12) $D_g{}^{air}$ = 1.7 . 10^{-5} . $1.03^{(T-293)}$

where T is the absolute temperature (K). The diffusion coefficient for air at 20 °C was estimated from data of Bruckler et al (7). Values of θ_v, θ_t and T are required as input variables for the model.

As mentioned before, the thickness of the top compartment L_1 is fixed for a volatilization event; the bottom of the second compartment is assumed to move with the concentration front of $NH_4{}^+$. The front will move downwards when precipitation exceeds evaporation (P>E); it will move upwards when E>P. The movement of the front J_s ($m.s^{-1}$) is derived from the water flux J_w following :

(13) $J_s = \dfrac{J_w}{\theta_v} . \dfrac{1}{1+R_D}$ with $R_D = \dfrac{D_b.[NH_4{}^+]_s}{\theta_v.[NH_4{}^+]_{aq}}$

where R_D is the retardation factor for the transport of $NH_4{}^+$, calculated as the ratio between the amount of $NH_4{}^+$ in the solid phase and the amount in the liquid phase (second compartment). D_b = dry bulk density ($kg.m^{-3}$), θ_v = water filled pore fraction, $[NH_4{}^+]_s$ = concentration on the solid phase ($\mu gN/kg$ dry soil), $[NH_4{}^+]_{aq}$ = concentration in solution ($\mu gN.m^{-3}$). The model calculates the thickness of the bottom compartment L_2, after each timestep from J_s and L_2', the thickness prior to the timestep, according to

(14) $L_2 = L_2' - J_s.DELT$

where DELT is the current timestep (s).

Biological processes like nitrification act as a sink for $NH_4{}^+$. An estimation of the amount nitrified ($\mu gN.m^{-2}$) on several moments during the volatilization event to be modelled is required as input for the model. The total rate of nitrification, s_t ($\mu gN.m^{-2}.s^{-1}$), calculated from this amount, is distributed over the compartments proportional to the distribution of NH_x

(15) $s_1 = s_t.NH_{x.1}/(NH_{x.1}+NH_{x.2})$ and $s_2 = s_t.NH_{x.2}/(NH_{x.1}+NH_{x.2})$

The amount of NH_x in the compartments can be calculated from the amount prior to the timestep, NH_x', the rate of volatilization R, convective transport J_c, diffusive transport J_d, and nitrification rate s_1 or s_2 :

(16) $NH_{x.1} = NH_{x.1}' + [J_c + J_d - s_1 - R]$. DELT
(17) $NH_{x.2} = NH_{x.2}' - [J_c + J_d + s_2]$. DELT

where J_c, J_d, s_1, s_2 and R are all expressed in ($\mu gN.m^{-2}.s^{-1}$)

From the total amount $NH_{x.1}$ and $NH_{x.2}$ in the compartments the distribution over the different species ($NH_4{}^+$ and NH_3) and phases (aqueous and gaseous) has to be calculated. $NH_{x.1}$ and $NH_{x.2}$ can be expressed in terms of the species, e.g. for compartment 1:

(18) $NH_{x,1} = L_1.\left[\theta_g.[NH_3]_{g.1} + \theta_v.[NH_3{}^0]_{aq.1} + \theta_v.[NH_4{}^+]_{aq,1} + D_b.[NH_4{}^+]_{s.1}\right]$

To solve $[NH_3]_g$ from this equation, $[NH_3]_g$ ($\mu g.m^{-3}$), $[NH_3{}^0]_{aq}$ ($\mu g.m^{-3}$) and $[NH_4{}^+]_s$ ($\mu g/kg$ dry soil) must be expressed in $[NH_4{}^+]_{aq}$, using the following relationships:

The partitioning of $NH_4{}^+$ between the solid phase and solution can be described according to either a linear isotherm, the Freundlich equation or the Langmuir equation :

	linear	Freundlich	Langmuir
(19) $[NH_4{}^+]_s$ =	$a.[NH_4{}^+]_{aq}$	$a.([NH_4{}^+]_{aq})^b$	$\dfrac{a.b.[NH_4{}^+]_{aq}}{1+b.[NH_4{}^+]_{aq}}$

The choice of the type of equation and the values of the parameters a and b are required as input for the model.

NH_4^+ and NH_3^0 are related following

(20) NH_4^+ <---> $NH_3^0 + H^+$ with a dissociation constant K_a :

(21) $K_a = \dfrac{[NH_3^0].[H^+]}{[NH_4^+]}$ or $[NH_3^0] = \dfrac{K_a.[NH_4^+]}{[H^+]} = \dfrac{K_a.[NH_4^+]}{10^{-pH}}$

where K_a is calculated from (see Hales & Drewes (8))

(22) $\log K_a = -0.09018 - 2729.92T^{-1}$

Data on soil pH and soil absolute temperature T are required as input for the model; See appendix for more details on pH.

The ratio between NH_3 in the soil solution and in the soil gas phase is calculated from Henry's law

(23) $K_h = [NH_3^0]_{aq}/[NH_3]_g$

where K_h is calculated from (see Hales and Drewes (8))

(24) $\log K_h = -1.69 + 1477.7T^{-1}$

When $[NH_3]_g$ is calculated, C_s from the flux equation (3) is known, and the transfer coefficient k must be calculated in the transfer module.

4. TRANSFER MODULE

Equation (3) gives the rate of volatilization when the ammonia profile in the air is in equilibrium with the concentration at the surface. The transfer coefficient k from this equation is defined by :

(25) $k = \dfrac{1}{r_a + r_b + r_s}$

where r_a, r_b and r_s (all in $(s.m^{-1})$) are the aerodynamic resistance, the resistance of the quasi-laminar interface and the surface resistance, respectively.

The aerodynamic resistance between the observation height and the surface depends on several factors: the roughness length z_0 (m) of the surface, which varies between 1 mm for smooth bare soil and 10 cm for a surface with tall vegetation, the friction velocity $u*$ ($m.s^{-1}$) calculated from the logarithmic wind profile, and the dimensions of the experimental plot (see ref. (4) for details). For arable soil and a circular plot with a radius of 21.25 m (which was used to test the model) r_a can be estimated as :

$r_a \approx 11.4/u*$

The resistance of the quasi-laminar interface represents the additional resistance for molecular diffusion through the interface. Hardly any information exists on values for NH_3, but calculations have been performed analogous to the behaviour of water and heat. It appears that r_b mainly depends on the friction velocity $u*$, and can be approximated with :

$r_b \approx 5.8/u*$

The surface (soil/slurry) resistance is the result of the diffusion process of ammoniacal-N species from inside the soil layer towards the air. Contrary to ref. (4) in the following also the diffusion of NH_4^+ was taken into account. The main parameter which determines the resistance is the distance over which diffusion has to take place, which in turn depends on the thickness of the surface layer for which calculations are performed. Further, the diffusion coefficients for NH_4^+ and NH_3^0 in the water phase (D_{aq}) and for NH_3 in the gas phase (D_g), the concentration of H^+ and the values of K_h and K_a (see eq. 23 and 21) influence the resistance, resulting in :

(26) $r_s = \dfrac{0.5\ L_1}{D_{aq}.K_h + D_{aq}.K_h.[H^+]/K_a + D_g}$

The factor 0.5 L_1 follows from the thickness of the top compartment L_1, and represents the mean distance for NH_3 to reach the soil surface. The terms in the denominator refer to diffusion of $NH_3{}^0$, $NH_4{}^+$ and NH_3 respectively. Summarizing, the transfer coefficient k can be calculated as :

$$(27)\quad k = \frac{1}{r_a + r_b + r_s} = \frac{1}{\dfrac{11.4}{u^*} + \dfrac{5.8}{u^*} + \dfrac{0.5\,L_1}{D_{aq}.K_h.(1+[H^+]/K_a) + D_g}}$$

From this equation it follows that when u^* increases (higher wind speed) both r_a and r_b decrease, so k will rise and thus volatilization; the rise of k has a limit when $(r_a+r_b) \ll r_s$, then $k \approx 1/r_s$. If temperature rises, both D_{aq} and D_g increase (see eq. 11 and 12); however, K_h decreases more strongly, and as a result k will decrease.

In figure 2 the dependence of k on u^* is shown, for different temperatures (0,4,8,12,16 and 20 °C, 2 mm) and values of L_1 (2,4 and 10 mm, 20 °C).

Fig. 2 Dependence of transfer coefficient k on friction velocity u^* at different values of L_1 and temperature

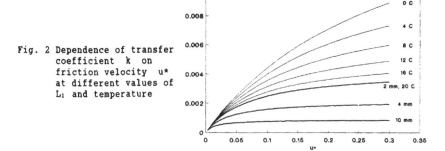

It appears that k diminishes strongly when L_1 increases, and also when temperature increases. At 20 °C, above a value of 0.20 u^* hardly influences the value of k, so k becomes independent of wind speed: the resistance of the soil against diffusive transport becomes determining. Calculations were performed for θ_t =0.6, water filled θ_v ≈0.25 and pH=8.0.

In figure 3 the dependence of k on the gas filled pore fraction θ_g is shown, for two values of u^*, where the value of 0.4 corresponds with the highest value found during the experiments, and 0.1 with a low wind speed. Calculations were performed for a temperature of 20 °C and for L_1 = 2 mm. The influence of θ_g on k follows from its influence on the gaseous diffusion coefficient D_g, according to eq. (10b). When u^* is low, soil resistance becomes less important, so the gas filled pore fraction has less influence on k. In this case $(r_a+r_b) \gg r_s$, so k has a maximum of $1/(r_a+r_b)$.

pH values were taken as 8.0 and 7.0; from eq. 26 it follows that at a lower pH the diffusion of $NH_4{}^+$ becomes more important when the soil is wet. It appears that k has a minimum; when θ_g increases, k increases due to diffusion of NH_3. When the soil is wet (low values of θ_g), k increases, especially due to diffusion of $NH_4{}^+$, which is most pronounced at lower values of pH. It has to be remembered, however, that the concentration of NH_3 in the soil gas phase, and thus the concentration gradient towards the atmosphere, increases with a factor 10 when pH changes from 7 to 8. The influence of pH on k partly compensates the effect of pH on $[NH_3]$.

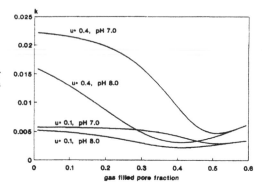

Fig. 3 Dependence of transfer
coefficient k on gas
filled pore fraction
(total porosity 0.6)

5. REQUIRED PARAMETERS

The following parameters are required as input for the model, and have to be
determined once :
* amount of ammoniacal N applied
* initial depth of placement of N
* adsorption parameters (distribution NH_4^+ soil/solution)
* pore volume
* bulk density
* surface roughness length and field size

The following parameters have to be determined frequently :
* u^* from wind speed
* soil temperature
* moisture content of soil
* pH of soil surface
* NH_3 background concentration
* rainfall rate
* (potential) evaporation rate
* loss of ammonium due to biological processes (nitrification)

6. FIELD EXPERIMENTS

In order to test the transfer model as described above four field experiments were carried out where emission of ammonia was measured after application of animal manure. Experimental setup and results are described in
detail elsewhere (3). In the next table some data on the experiments are
given for the first 7 days. The manure was applied using a computerized
slurry-spreader; in exp. 3 and 4 the slurry was mixed through the upper 6 cm
of the soil, immediately after application, using a cultivator.
In the table temp. is the mean soil temperature at -0.02 m ($^\circ$C); wind
is mean wind speed at 10.0 m (m.s^{-1}), rain is cumulative amount (mm), radiation is mean of global radiation (J.cm^{-2}.hr^{-1}), and flux is percentage of
NH_x applied volatilized during the period.

exp.	time	way of application	soil temp.	wind speed	rain	global radiation	flux %
1	sept. 1987	surface	10.1	3.9	3.1	1054	65
2	nov. 1987	surface	9.7	2.2	2.0	182	30
3	april 1988	mixed	8.7	3.6	2.2	1434	15
4	may 1988	mixed	9.6	3.8	3.3	1358	10

In fig. 4 the fluxes are compared; the difference between exp. 1 and 2 is remarkable. In exp.1 a rapid volatilization occured during the first two days, when mean soil temperature was 12 and 10 °C, respectively; in exp.2 on these days mean temperature was only 7 and 8 °C. Mean global radiation was much larger in exp.1, but volatilization does not seem to be directly proportional to radiation, at short notice.

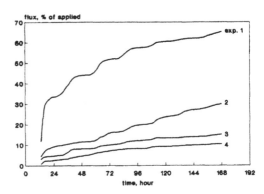

Fig. 4 Volatilization as a
        percentage of NHx
        applied in 4 field
        experiments
        1,2 surface applied
        (2 low radiation)
        3,4 mixed through 6 cm

7. CONCLUSIONS

Volatilization highly depends on
- wind speed
- soil temperature
- evaporation / rainfall
- gas filled pore fraction (dryness)

These factors depend on weather conditions, which cannot be predicted.

Another key factor is soil pH, which determines the equilibrium between NH_4^+ and NH_3^0 according to eq. (20). Soil pH is influenced by a number of processes (see appendix); this makes it difficult to predict soil pH in a model.

Therefore, a model like the one presented here can not predict what will happen (volatilize) after surface application of manure, but may help to understand what happens.

8. REFERENCES

(1) Beauchamp, E.G., G.E. Kidd & G.W. Thurtell, 1982. Ammonia volatilization from liquid dairy cattle manure in the field. Can. J. Soil Sci. 62: 11-19.

(2) Molen, J. van der, D.W. Bussink, N. Vertregt, H.G. van Faassen & D.J. den Boer, 1989. Ammonia volatilization from arable and grassland soils. In : K. Henriksen (Ed.), Nitrogen in Organic Wastes Applied to Soils. Academic Press, London, 1989. p. 185-201.

(3) Molen, J. van der, H.G. van Faassen, M.Y. Leclerc & R. Vriesema, 1990. Ammonia volatilization from arable land after application of cattle slurry; 1. Field estimates. Neth. Journal of Agric. Sci. (in press).

(4) Molen, J. van der, A.C.M. Beljaars, W.J. Chardon, W.A. Jury & H.G. van Faassen, 1990. Ammonia volatilization from arable land after application of cattle slurry; 2. Derivation of a transfer model. Neth. Journal of Agric. Sci. (in press).

(5) Jury, W.A., W.F. Spencer & W.J. Farmer, 1983. Behavior assessment model for trace organics in soil I. Model description. J. Environ. Qual. 12: 558-564.

(6) Yuan-Hui, L. & S. Gregory, 1974. Diffusion of ions in sea water and in deep-sea sediments. Geoch. Cosmoch. Acta 38: 703-714.

(7) Bruckler, L., B.C. Ball & P. Renault, 1989. Laboratory estimation of gas diffusion coefficient and effective porosity in soils. J. Soil Sci. 147: 1-10.

(8) Hales, J.M. & D.R. Drewes, 1979. Solubility of ammonia at low concentrations. Atmospheric Environment 13: 1133-1147.

9. APPENDIX on pH

1. Calculation of average pH

The pH is used in the model for the calculation of the concentration of NH_3, following :

$$NH_4^+ \; \langle -- \rangle \; NH_3 + H^+$$

with a dissociation constant K_a :

$$K_a = \frac{[NH_3][H^+]}{[NH_4^+]}$$

or : $[NH_3] = K_a \cdot [NH_4^+] \cdot [H^+]^{-1}$

or : $[NH_3] = K_a \cdot [NH_4^+] \cdot 10^{pH}$

When the value of pH shows a spatial variation in the field, an average value has to be calculated. This can be done by simple mathemathics on pH, but this gives an incorrect result; in fact the value of 10^{pH} is used to calculate $[NH_3]$, so the average of values of 10^{pH} has to be used. The difference between both methods is shown in the following example :

				mean pH	mean 10^{pH}	ratio
pH	6	7	8	7	10^7	1
10^{pH}	10^6	10^7	10^8		$10^{7.6}$	3.7

The consequence of the second method is that a 3.7 x higher value of $[NH_3]$ is calculated.

2. Processes in soil which influence pH.

After introduction of manure into a soil the following reactions, by which pH is influenced may occur :

a. Hydrolysis of urea (mainly already during storage) :
   $$CO(NH_2)_2 + 3H_2O \longrightarrow (NH_4)_2CO_3 + H_2O \longrightarrow 2NH_4^+ + HCO_3^- + OH^-$$
b. Dissociation of ammonium, volatilization of ammonia :
   $$NH_4^+ \longleftrightarrow NH_3^0 + H^+ \quad \text{followed by :} \quad NH_3^0 \longleftrightarrow NH_3^0 (g)$$
c. Association of (bi)carbonate, volatilization of CO_2 :
   $$CO_3^{2-} + 2 H^+ \longleftrightarrow HCO_3^- + H^+ \longleftrightarrow H_2CO_3^0 \longleftrightarrow CO_2(g) + H_2O$$
d. Plant uptake of NH_4^+ :
   $$plant\text{-}H^+ + NH_4^+ \longrightarrow plant\text{-}NH_4^+ + H^+$$
e. Conversion of NH_4^+ into NO_3^- (with NO_2^- as intermediate product) :
   $$NH_4^+ + 2 O_2 \longrightarrow NO_3^- + 2 H^+ + H_2O$$
f. Plant uptake of nitrate :
   $$plant\text{-}OH^- + NO_3^- \longrightarrow plant\text{-}NO_3^- + OH^-$$
g. Denitrification of nitrate into N_2 under anaerobic conditions :
   $$10[H] + 2 H^+ + 2 NO_3^- \longrightarrow N_2 (g) + 6 H_2O$$
h. or, when less organic substrate is available, into N_2O :
   $$8[H] + 2 H^+ + 2 NO_3^- \longrightarrow N_2O(g) + 5 H_2O$$
i. Precipitation/dissolution of Ca- or Mg-carbonate, where Ca^{2+} or Mg^{2+} from the soil is exchanged by NH_4^+ and K^+ from the manure :
   $$HCO_3^- + Ca^{2+} \longleftrightarrow H^+ + CaCO_3$$
j. Breakdown of acetic acid or other organic acids :
   $$CH_3COOH + 2 O_2 \longrightarrow H_2CO_3 + HCO_3^- + H^+$$
k. Exchange of cations with H^+ from soil organic matter, e.g. for NH_4^+ :
   $$SOH + NH_4^+ \longleftrightarrow SO\text{-}NH_4 + H^+$$

The net effect on pH of reactions where CO_2 is involved (a,d,f,j) strongly depends on the carbonate equilibrium (eq. c). It follows from the equations above that a prediction of the change in time of the pH of the soil/manure will be complicated.

POLICY FOR AMMONIA EMISSIONS IN IRELAND

O.T. CARTON and H. TUNNEY

Johnstown Castle Research Centre,
Wexford,
Ireland.

Summary
     The environmental implications of ammonia emissions are cause for
serious national concern. Ammonia emissions from livestock production
in Ireland has been estimated at almost 110 kilo tonnes per annum.
This is equivalent to 0.30 of national fertiliser nitrogen usage.
Consequently reducing ammonia volatilisation from agriculture is a
national priority.

1. INTRODUCTION

In Ireland, as well as the other European countries, agriculture, is the most
important source of ammonia emission (Buijsman et al., 1987). The sources
include the excreta of the grazing animals, the landspreading of animal
wastes, the farm buildings and slurry stores and the spreading of fertiliser
nitrogen. The loss of nitrogen, as ammonia, represents a significant loss to
the farmer of a potentially useful and expensive nutrient. Ammonia in the
atmosphere results in acid rain and the deposition of sulphate.

In this paper an attempt is made to quantify total ammonia emission from
Irish agriculture. A brief description is also given of the current research
programme to minimize ammonia emission from farming practice in Ireland.

2. Ammonia emission from agriculture in Ireland

In Ireland annual fresh manure production has been estimated at almost 87
million tonnes (Table 1). Eighty five million tonnes from grazing animals
(bovines and ovines) and 2 million tonnes from pigs and poultry. During the
winter period bovines are housed for approximately 135 days while ovines are
indoors only for short periods, generally at lambing time. This indoor winter
period for grazing animals and the manure from pig and poultry production
results in 29 million tonnes of manure annually. This manure is land spread.
As 0.93 of our total land area is grassland this manure is applied almost
totally to grass. The remaining manure, 58 million tonnes, is automatically
recycled during grazing.

Livestock Type	Numbers ('000's)	Annual Waste Production (kilo ton, fresh)
Bovines	6,800	76,588
Ovines	7,729	8,288
Pigs	984	1,804
Poultry	8,300	227
Total		86,907

Table 1: Estimated annual waste production from livestock in Ireland

Using published data an attempt has been made to quantify the ammonia
emissions from livestock production in Ireland. The results are summarised
in Table 2. The estimate shows little change from that calculated by Buijsman
et al., (1987) based on animal numbers in 1980. Grazing animals,
landspreading of manures and emissions from stored manures account for 0.48,

0.40 and 0.12, respectively, of total ammonia emissions. Current, annual fertiliser nitrogen use in Ireland is almost 370,000 tonnes. Therefore, ammonia emissions from livestock production are equivalent to 0.30 of our fertiliser nitrogen usage. Nitrogen currently costs the farmer 533 ECU per tonne. Therefore, ammonia emissions are worth 57.7 million ECU annually. This simple calculation takes no account of the indirect costs imposed by the environmental pollution from ammonia emission to the atmosphere.

	Storage	Spreading	Grazing	Total
Bovines	11.33	40.79	43.68	95.80
Ovines	0.02	0.06	8.52	8.60
Pigs	0.78	1.40	-	2.18
Poultry	0.90	0.72	-	1.62
Total	13.03	42.97	52.2	108.20

Table 2. Estimate ammonia emissions from livestock production (kilo ton)

Consequently, one of the primary objectives of the environmental research programme at Johnstown Castle is a reduction in the ammonia emissions from livestock production. Principally, work is focused on reducing emissions form landspreading of manures. Associated with this objective is the requirement to reduce the odour emission from landspreading. Work is also concerned with volatilization of ammonia from the landspreading of fertiliser nitrogen.

3. Reducing Ammonia Emission from Landspreading of Manure

Initially, deep (100 mm) soil injection as a means of reducing ammonia emission from the landspreading of manures was evaluated. Tunney and Molloy (1985) concluded that soil injection did reduce ammonia emissions but the crop recovery of the ammonia nitrogen from the manures remained poor. This may have resulted from denitrification losses in the deep slits. Thompson et al., (1987) reported an increase in denitrification losses from deep injection compared with surface applied slurry. More recently, Long and Gracey (1990) concluded deep injection, in summer, can be an effective means of utilizing slurry nitrogen in terms of herbage dry matter production. However, it is unlikely that deep injection of manures will be widely adopted on Irish farms as farm size is generally small (less than 0.10 of farms having more than 50 hectares) and almost 0.40 of our soils are classified as wet.

The current programme is examining the potential of shallow injection as a means of reducing ammonia emissions and improving crop responses to the nitrogen from landspread manure. Results from Kiely (1988) indicated significant improvements in grass dry matter response to shallow injected compared with surface spread manure. Carney and Dodd (1989) reported a 0.70 reduction in odour emission from shallow injection compared with surface spreading. Bandspreading, the placing of slurry in rows 30 to 50 mm wide on the soil surface at 300 mm intervals, was also found to be superior to the splashplate method of application in terms of crop dry matter responses (Kiely, 1988). In contrast Pain (1988) on small plots reported no difference in total ammonia emission from bandspread or splashplate spread slurry. However, they did find that emissions within 24 hours of spreading were lower from the bandspread slurry. This apparent discrepancy requires clarification. To date no quantification of odour emission from bandspreading has been made under field conditions. However, bandspreading has the advantages of being a relatively cheap option requiring very little additional equipment.

4. Conclusions.

Ammonia emissions from Irish livestock production systems is resulting in environmental damage and a significant loss of potentially utilisable nitrogen. Work will continue on spreading techniques with or without some slurry treatments as means of reducing ammonia emission from the landspreading of animal wastes. Increasingly in Ireland there is a growing public intolerance of odour emissions from the landspreading of animal wastes. Though odour emission will not be measured it is assumed that a reduction in ammonia emission will be accompanied by a reduction in odour emission.

REFERENCES

(1)   BUIJSMAN, ED., MAAR, HAGS F. and ADMAN, W.A. (1987). Anthropogenic NH_3 emissions in Europe. Atmospheric Environment 21, 1009-1022.

(2)   TUNNEY, H. and MOLLOY, S.P. (1985). Comparison of grass production with soil injected and surface spread cattle slurry. In: Efficient land use of sludge and manure. (ed. A. Damkofoed, J.H. Williams and P. L'Hermite). Published by Elsevier Applied Science Publishers 90-98.

(3)   THOMPSON, R.B., RYDEN, J.C. and LOCKYER, D.R. (1987). Fate of nitrogen in cattle slurry following surface application or injection to grassland. Journal of Soil Science, 28, 689-700.

(4)   LONG, F.N.J. and GRACEY, H.I. (1990). Herbage production and nitrogen recovery from slurry injection and fertiliser nitrogen application. Grass and Forage Science, 45, 77-82.

(5)   KIELY, P.J. (1988). The effect of spreading method on slurry N utilization by grassland. In: Proceedings of the 12th General Meeting of the European Grassland Federation, Dublin, 353-357.

(6)   CARNEY, P.G. and DODD, V.A. (1989). The measurement of agricultural malodours. Journal of Agricultural Engineering Research, 43, 197-209.

(8)   PAIN, B.F. (1988). Ammonia losses during and following the application of slurry to land. In: Minutes of a Joint Meeting of the FAO - European Network on 'Animal Waste Utilization' and Working Parties 4 and 5 of Cost Project 681. (ed. H. Vetter, G. Steffens and P. L'Hermite). Publish by the commission of the European Communities.

Session V

AMMONIA AND ODOUR LOSSES FROM GRASSLAND

Chairperson : J. HARTUNG

ODOUR AND AMMONIA EMISSIONS FROM GRASSLAND AND ARABLE LAND

A. KLASINK, G. STEFFENS and H.-H. KOWALEWSKY

Landwirtschaftskammer Weser-Ems,
Landwirtschaftliche Untersuchungs- und Forschungsanstalt,
Mars-la-Tour-Str. 4, D-2900 Oldenburg, BRD

Summary
        Open slurry tanks emit intensive odours, but odours from slurried fields reach higher distances. Ammonia is not the main odour component, numerous factors influence odour intensity. CaO added to cattle slurry caused increased NH_3 emission and decreased odour intensity. Nitric acid and superphosphate didn't decrease odour emission, but NH_3 emission. Small amounts of CaO didn't cause higher NH_3 loss than untreated slurry. Ammonia loss following land spreading of pig slurry in January and February depends on air temperature, rainfall and soil humidity.

1. INTRODUCTION

In future land spreading of slurry and manure will only be acceptable if we will succeed in diminishing odour and ammonia emissions. Therefore the efforts measuring odour and ammonia emissions have been intensified. But not only characterization of odour and quantification of ammonia loss under different conditions is necessary, we need also research activities finding methods which result in diminishing odour emission and ammonia loss under field conditions. In combination with a research project on better use of slurry we are carrying out a small programme concerning odour and ammonia emission. First results will be presented here.

2. METHODS

We measure ammonia loss following the application of pig slurry to land with the wind tunnels system. On small plots (1.2 m²) the ammonia emission rate has been measured with time discontinuously round about 24 h. The product of air volume and ammonia concentration (μ NH_3/m^3) gives the amount of ammonia loss (kg/ha).

In another experiment cattle slurry was filled in containers (140 l) and then stirred while adding nitric acid or CaO or superphosphate. After application of the agents and stirring 2-3 minutes ammonia volatilization and odour has been measured. This procedure has been repeated two or three weeks after storing the treated slurry in the covered containers.

The aim of adding CaO was mainly to get better homogenized slurry for better distribution. The question was, if the ammonia loss was nevertheless high.

After three weeks storage the ammonia loss has been measured following the application of slurry to grassland with the wind tunnels system.

3. RESULTS
3.1 Former results

Kowalewsky (1981) compared sensoric odour impressions with odour intensity based on chemical analysis of different components (NH_3, H_2S, acetic acid e.a.). His investigations have shown that odour intensity at the emission source is especially high at open slurry tanks, but that odours from slurried fields reach much higher distances. Pig and poultry stables are less severe in this regard. Ammonia emission is more intensive from field application than from stables or from containers (Figure 1). This shows that ammonia is not the main component of odour.

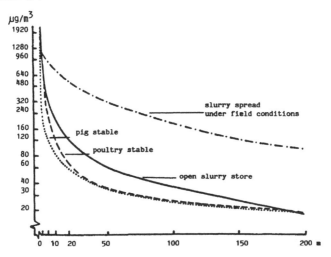

Figure 1:   Ammonia content of air in different distances from different emission sources

Ammonia losses depend on numerous factors which influence the emission rate under field conditions (Table 1). In combination with our field research programme we will try to quantify the influence of these factors on ammonia volatilization.

o   time of spreading concerning climate factors
     (temperature, windspeed, rainfall)
o   soil type
o   slurry type (cattle, pig, poultry)
o   dry matter content of slurry
o   kind of spreading
o   soil with and without plant growth

Table 1:   Parameters influencing ammonia volatilization under field conditions

3.2 Odour and ammonia emissions after adding agents to cattle slurry

Increasing amounts of CaO caused increasing pH, decreasing odour emission and increasing NH_3 emission (Table 2). Three weeks later the pH value didn't change, but the odour units* (GE/m^3) decreased strongly at the treatment with 5.0 and 7.5 kg CaO/m^3 cattle slurry. The NH_3 emission was nearly as high as three weeks before. We see, that there is no positive connection between odour emission and NH_3 emission. Surely odour emission depends mainly on other components (organic acids, H_2S, phenol, p-cresol etc.). The increasing pH changes the microbiological situation of slurry. This may be the reason of diminished odour emission by adding CaO.

Treatment	pH 13.2.90	pH 7.3.90	odour units, GE/m^3 12.2.90	6.3.90	NH_3, $\mu g/m^3$ 16.2.90	5.3.90
1. without CaO	7.25	7.37	26789	27552	400	1172
2. 2.5 kg CaO/m^3	7.93	7.73	13880	29184	3349	2856
3. 5.0 kg CaO/m^3	8.84	8.82	5316	4864	35226	25923
4. 7.5 kg CaO/m^3	9.87	9.82	3884	640	141290	134483

Table 2: pH, odour units and NH_3-concentration after adding CaO to cattle slurry

The addition of superphosphate had no influence upon pH, but nitric acid decreased pH (Table 3). Odour control after stirring the slurry lead to the result, that odour development was more intensive after treatment with superphosphate and especially high after treatment with nitric acid. About three weeks later a further odour control resulted in a higher odour emission of the untreated slurry compared with the treated slurry. In spite of high odour emission the ammonia emission decreased most after treatment with nitric acid.

Treatment	pH 13.2.90	pH 7.3.90	odour units, GE/m^3 16.2.90	6.3.90	NH_3, $\mu g/m^3$ 16.2.90	14.3.90
1. control	7.25	7.37	38976	65536	400	1842
2. 5 kg super-phosphate/ha	7.00	7.40	56736	27592	270	2097
3. 10 l HNO_3/m^3	–	4.39	123776	45040	129	142

Table 3: pH, odour units and NH_3-concentration after adding superphosphate or nitric acid to cattle slurry

---

* Olfactometry, Odour Threshold Determination, Fundamentals VDI 3881

Three weeks after treatment we measured ammonia loss after slurry spreading on grassland (Table 4). In all cases ammonia loss was highest during the first hour after application. Increasing CaO content raised in this time the ammonia loss. In the following hours the relations between the different treatments changed: untreated slurry lost more ammonia then the CaO-treated plots. At the end there was only a small difference in total ammonia loss between the treatments except the highest amount of added CaO. The total loss ranged between 26.1% (lowest amount of CaO) and 45.0 % (highest amount of CaO) of total ammonia content.

Treatment	after			kg/ha total loss	total loss in % of	
	1h	2h	3h		total N	NH_3-N
		kg/ha				
1. Without CaO	7.6	3.7	3.1	14.4	19.5	32.7
2. 2.5 kg CaO/m³	6.0	3.0	2.5	11.5	15.5	26.1
3. 5.0 kg CaO/m³	8.2	2.6	2.5	13.3	18.0	30.2
4. 7.5 kg CaO/m³	16.6	1.8	1.5	19.8	26.8	45.0
5. 5 kg superphosphate/m³	7.3	3.1	2.9	13.3	18.0	30.2

Amount of application = 20 m³/ha cattle slurry
d.m. 7.9 %, 74 kg/ha total N, 44 kg/ha NH_3-N

Table 4: Loss of ammonia following the application of cattle slurry to grassland

May be that CaO improves slurry distribution and for instance 5 kg CaO/m³ reduce odour emission without increasing the ammonia loss. But further research is necessary.

3.3 Ammonia loss under field conditions

In the north-west region of Germany farmers are permitted to start slurry application on arable land and grassland in February. We try to find out the conditions of ammonia volatilization depending especially on time of spreading, soil type and further parameters (s. Table 1). The first application of 20 m³/ha pig slurry on arable land (rye, sandy soil) was in January. Until now we have registrated ammonia loss at three terms. The programme will be continued.

In agreement with results of other authors we found the highest rates of ammonia volatilization during the first hours (Figures 2, 3 and 4). Under the same field conditions the ammonia loss from sandy soil was higher than from clay soil. About twenty four hours after application the ammonia rate in the air was near to the untreated plot.

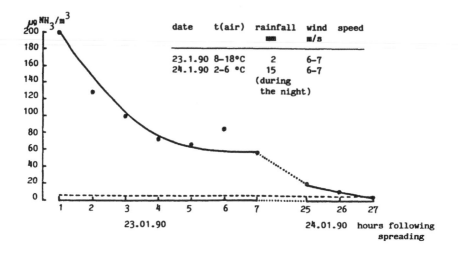

Figure 2:   Ammonia loss following land spreading of pig slurry
              Soil type = hS

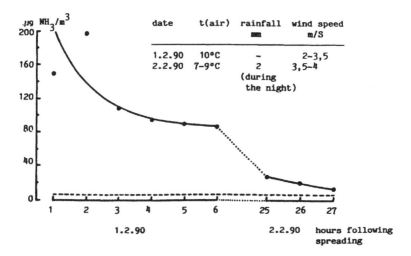

Figure 3:   Ammonia loss following land spreading of pig slurry
              Soil type = hS

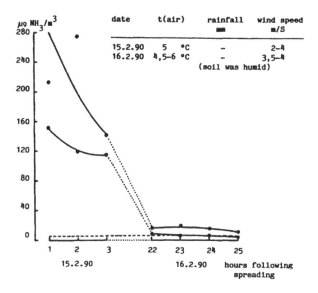

Figure 4:   Ammonia loss following land spreading of pig slurry
              Soil type = hS and uT

The different application dates show the dependence of ammonia volatilization on weather conditions (Table 5). The calculated total ammonia loss is ranging between 8.1 % (clay soil, lower temperature, humid soil, see figure 4) and 40.3 % (sandy soil, higher temperature, no rainfall, see figure 3). These few results demonstrate the high importance of soil type, air temperature, rainfall or soil humidity on ammonia volatilization under field conditions on arable land. The investigations concerning ammonia volatilization will be continued in our institute.

Treatment	Application rate, kg		Ammonia loss		
	total N	NH_3–N	kg NH^3–N	% of total N	% of NH_3–N
23./24.1.90	138	84	18.4	13.3	21.9
1./2.2.90	144	88	35.5	24.7	40.3
15./16.2.90 (hS)	128	78	17.4	13.6	22.3
15./16.2.90 (uT)	128	78	6.3	4.9	8.1

Table 5:   Ammonia loss at different application dates under different climate
             conditions

REFERENCES

(1)   Anonymous, Emissionen von Ammoniak. – Quellen – Verbleib – Wirkungen – Schutzmaßnahmen –. Arbeitsmaterialien des Bundesamtes für Ernährung und Forstwirtschaft, Frankfurt/Main, Juni 1989.

(2)   BLESS, H.G., Ammoniak-Emissionen nach der Ausbringung von Flüssigmist. Studie im Auftrag des Ministers für Ernährung, Landwirtschaft, Forsten und Fischerei des Landes Schleswig–Holstein und der Landwirtschaftskammer Schleswig–Holstein, Kiel, Januar 1990.

(3)   KOWALEWSKY, H.-H., Messen und Bewerten von Geruchsemissionen. KTBL–Schrift 260, 1981, Landwirtschaftsverlag GmbH, 4400 Münster–Hiltrup (Westf.).

(4)   NIELSEN, V.C., VOORBURG, J.H. and P. L'HERMITE (Editors). Odour prevention and control of organic sludge and livestock farming. Commission of the European Communities. Proceedings of a seminar held in Silsoe, UK, 15–19 April 1985. Elsevier Applied Science Publishers, London and New York, 1986.

(5)   VETTER, H., STEFFENS, G. and P. L'HERMITE (Editors), Safe and efficient slurry utilization. Concerted Action – Treatment and use of organic sludge and liquid agricultural wastes. Cost project 681, Liebefeld (CH), 20 – 21 June 1988.

AMMONIA EMISSIONS FROM GRAZING

N. VERTREGT and B. RUTGERS

Centre for Agrobiological Research
Bornsesteeg 65, 6700 AA Wageningen, The Netherlands

Summary
        The ammonia-N emissions from artificially prepared urine patches on grassland on a sandy soil were measured with a windtunnel method. Artificial urine with a N-content of 6-12 $g.l^{-1}$ was applied at a rate of 5 $l.m^{-2}$. At a urine application rate of 600 kg $N.ha^{-1}$ the ammonia-N emission varied between 6 % and 19 % (mean 13 %) of the urine-N.
It was argued that the emission rate is proportional to the urine-N concentration. The ammonia emission rate is 10 % at an average urine-N concentration of 9 $g.l^{-1}$. The results of the emission measurements were confirmed by N-budget analyses of the urine patch system.
        The total ammonia emissions from grazed pastures at various levels of N-supply were calculated using known data on grass production, intake and digestion. The calculated ammonia-N emissions varied from 16 kg $N.ha^{-1}$ at a N supply of 100 $kg.ha^{-1}$ to 38 $kg.ha^{-1}$ at a N-supply of 500 $kg.ha^{-1}$. The calculated ammonia emissions were similar to the results of measurements in grazed fields with the micrometeorological mass balance method. At low levels of N-supply the mass balance data were disproportionally low, probably as a result of lower urine-N concentration during grazing and an underestimation of the emission by the mass balance method due to a relatively high background interference.

1. INTRODUCTION
        Several years ago it was recognized that the ammonia emissions from animal excretion contribute significantly to the eutrophication and acidification of the natural environment.
About 60 % of the ammonia emission in the Netherlands is attributed to cattle and 60 % of the N-excretion of cattle occurs during the grazing period. For this reason the ammonia emission during grazing was studied in some detail. Two approaches were used in The Netherlands. In the CABO experiments use was made of a windtunnel placed over artificial urine patches on 1 m^2 plots of grassland (1). The NMI measured ammonia volatilization from grazed pastures with a micrometeorological mass balance method (2). When properly applied both methods give equal results (3).
        The ventilation rate in the tunnel has to be sufficiently high to ascertain climatic conditions in the tunnel similar to ambient conditions. The accuracy of the two methods will be discussed and the total ammonia emission from grazed pastures will be calculated.
The work was part of the Dutch Priority Program on Acidification.

2. EXPERIMENTS
        Windtunnel experiments were carried out in 1986 and 1987 on grassland on a sandy soil at the experimental farm Droevendaal at Wageningen. Air was filtered ammonia-free by forcing it through an airconditioning filter impregnated with orthophosphoric acid. The ammonia-free air was ventilated over the plots prepared with artificial urine according the Doak (4). The

urine application rate was 600 kg N.ha^{-1}, 5 liter per m^2 of urine with a concentration of 12 g N.l^{-1}. The ventilation rate was 1200 m^3 per hour, or 1.5 meter per second through a tunnel of 2 meter length, 0.55 meter width and 0.60 meter height. The air was continuously sampled after passage through the tunnel. The ammonia emission was calculated by multiplying the quantity of ammonia absorbed from the sampled air with the ratio of the measured volume of the ventilation air and the measured volume of the sampled air.

3. RESULTS AND DISCUSSION

Typical course lines for the ammonia emission and the pH of the soil surface are presented in Figure 1. The pH of the soil increases from 6 to 9 within a day after urine application. The pH decreases slowly to the initial value in a period of about 20 days. The ammonia volatilization rate declines rapidly during the first days after urine application.

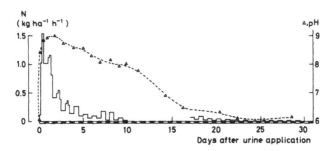

Fig. 1. Ammonia volatilization, kg N.ha^{-1}h^{-1} and soil surface pH following a urine-N application of 600 kg.ha^{-1}.

The total ammonia-N emissions over periods of 10 days are represented in figure 2 for 17 experiments on sandy soil with a urine-N application rate of approximately 600 kg per ha. The total N emission varied between 23 and 98 kg N.ha^{-1}. No definite relations between temperature, global radiation, soil moisture content or pH of the soil surface and the ammonia emission could be established.

Obviously the ammonia volatilization depends on more factors than the factors mentioned. Part of the emission is governed by physical and chemical reactions in the upper soil layer. Part of the urine adheres to the leaf and litter layer and will volatilize following different kinetics. As the total emission is relatively small, a substantial fraction will volatilize from the leaf and litter layer.
The volatilization from the top soil layer depends on the ammonium concentration in this layer. It is to be expected that the ammonium transport in the upper soil layers is influenced by the soil moisture content and the water supply to the soil. However, as stated before, no clear relation between soil moisture content and volatilization level was found.

The average emission during the first 10 days after urine application amounts to 10 % of the urine-N applied. The daily emission decreased to zero a month after that application. Consequently the total ammonia emission in a period of a month is 13 % of the urine-N.

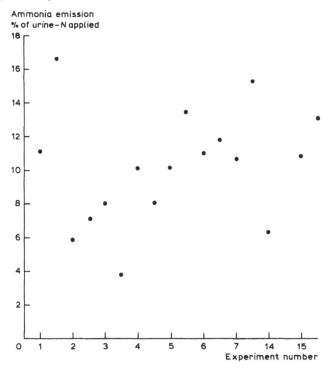

Fig. 2. Ammonia-N emission in % of N applied for successive identical experiments on sandy soil.

This value for the ammonia emission however, seems to be an overestimation of the true value of the emission under field conditions according to the following reasons. During the experiment it was recognized that the average N-concentration in the urine fluctuates around 9 g per liter. In the experiments artifical urine with a N-concentration of 12 g per liter was used. It is assumed that at least the relatively high ammonia emission during the first days after urine application is proportional to the urine-N concentration as this emission originates mainly from the urine adhering to the leaf and litter layer (6). Therefore the average ammonia emission of urine with a N-concentration of 9 g per liter can be estimated at 11 % of the urine-N application.
As mentioned before, the effect of soil temperature and soil moisture content on the ammonia emission could not be established from the

experiments. However, on the average the temperature under the tunnel was about 1 °C higher than outside it. The soil moisture content decreased gradually during the experimental period. As a consequence the measured ammonia emission in the tunnel system is at least theoretically somewhat higher than under field conditions. This effect is only temporary, as the volatilization is restricted by the ammonium transport to the topsoil layer. The correction for the slightly deviating temperature and soil moisture content on the ammonia volatilization rate has been estimated at 10 % (5).

The totalized corrections reduce the estimated average ammonia emission to 10 % of the urine-N at an urine-N concentration of 9 g.l^{-1}.

The ammonia emission from grazed pastures was measured directly with the micrometeorological mass balance method on plots that were fertilized with 550 kg N.ha^{-1}year^{-1} (2). The plots were grazed during one and a half day with a stocking density of 60 to 120 cows per hectare. The mean ammonia emission was determined to be 8 % of the estimated excreted N. The true ammonia emission may be underestimated by the mass balance method.

The first reason for this statement is that the limits of the measuring system are not exactly defined. Some of the volatilized ammonia may escape beyond the assumed limits of the ammonia profile.

Secondly the background concentration measured at the windward side of the field or at a high position on the central mast may be overestimated i.e. because of back diffusion. Consequently the calculated net emission has a relatively low value.

Thirdly with the mass balance method the net emission is measured. This net emission is the difference between the ammonia emission and the ammonia deposition from the atmosphere. As the windtunnels are ventilated with ammoniafree air this third factor is absent in the windtunnel method. In the Netherlands the atmospheric deposition can be estimated on 0,5 kg per hectare in a 10-day period. It can roughly be estimated that with the factors mentioned the ammonia emission as measured by the mass balance method has to be corrected from 8 % to 10 % of the excreted N.

It must be borne in mind that the variation in ammonia emission is considerable. However, both methods agree on the fact that 90 % (85-95 %) of the urine-N does not volatilize from urine patches.

To validate the data on ammonia emission we studied the conversion of the urine-N in urine patches in more detail.

4. N-BUDGET OF URINE PATCHES

Before the start of the experiments and ten days after urine application, soil samples were taken to a sufficient depth and analysed for the ammonium and nitrate content (6). At the same time the grass was sampled and analyzed. In many experiments the N-uptake by the grass was negligible, on the average 10 kg N.ha^{-1} in a 10-day period, partly because of growth inhibition due to the urine application. For all experiments on sandy soils in which 600 kg N.ha^{-1} was applied the relevant N-fractions at the 10th day after urine application are plotted in Figure 3. For each experiment the remaining part of the urine-ammonium (urea)-N is plotted on the abscissus against the amounts of volatilized ammonia-N, nitrate-N and N not accounted for (N_{naf}) on the ordinate. The calculated regression lines for NH_3, $NH_3+NO_3^-$ and $NH_3+NO_3^-$ + N_{naf} against the soil ammonium-N are drawn in figure 3. In a number of experiments the budget loss, N_{naf}, was relatively high. Theoretically, the budget loss could be attributed to inaccurate sampling methods. In some experiments a nitrification inhibitor was applied in order to prevent a rapid decrease in pH. In these experiments the final ammonium content of the soil was high, the

nitrification and the N-budget loss were both very low. So it was
concluded that the sampling procedure was correct.

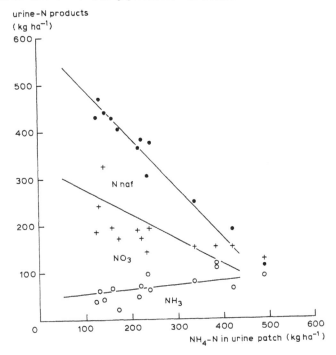

Fig. 3. N-budget of a urine patch, 10 days after application of 600 kg
urine-N ha^{-1}. x-axis: NH_4^+-N in soil. y-axis: NH_3-N volatilized
(o), NH_3-N + NO_3^--N in soil (+), NH_3-N + NO_3-N + N not accounted
for (•).

     In experiments with a low final ammonia content in the soil, the
nitrate content is high, the ammonia volatilization is relatively low and
the amount of N not accounted for is high. The ammonia volatilization is
high under conditions of low nitrification and high pH, but the increase
in ammonia emission is not proportional to the soil ammonium content.
Summarizing the ammonia volatilization is not only restricted by the
nitrification but also by interactions between ammonium and the soil
complex. The N-budget loss is coupled with the nitrification. Obviously a
large fraction of the urine-N gets lost during nitrification by a chemo-
denitrification process.

5. CALCULATION OF THE AMMONIA EMISSION FROM GRAZED PASTURES AT DIFFERENT
   RATES OF FERTILIZATION
     The yield and nitrogen content of grass in relation to the N
fertilization is calculated from a large number of measurements on sandy
soils in the Netherlands (7).

The grass intake of dairy cows is assumed to be 14.5 kg dry matter per day. The calculated dry matter and N intakes are plotted in quadrant I of Figure 4. The milk production is 20 kg per day with a nitrogen content of 0.53 % (8). The urine and faeces excretions are calculated using the digestion coefficients published in (9) and the milk production. The calculated values for the urine and faeces excretions are plotted in quadrant II. The ammonia volatilization emission is fixed at 10 % of the N-excretion. The calculated emissions are plotted in quadrant III and IV. By reducing the fertilization from 400 to 200 kg N per ha.year the ammonia emission decreases from 33 to 21 kg per hectare per year. The stocking density decreases at the same time from 3.9 to 3.1 dairy cows per hectare.

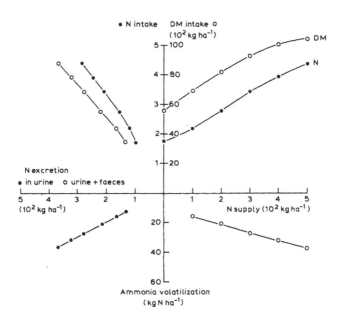

Fig. 4. Effect of N supply on the dry matter (o) and N (•) intake during grazing, on the N-excretion in urine (•) and urine + faeces (o) and the resulting ammonia-N volatilization assuming 10 % volatilization from excreted N in relation to N-excretion (•) and N-supply (o).

For swards receiving 400 kg N.ha^{-1} the calculated ammonia emissions are similar to the results of field measurements with the micrometeorological mass balance method (10). With lower N-supplies the ammonia volatilization as measured with the mass balance method decreases from 10 % to approximately 5 % of the N returned to the sward. This discrepancy between the results of the mass balance and the windtunnel methods can be partly explained by the estimated 30 % lower urine-N concentrations during grazing in low-N swards, as the relative ammonia emission is proportional

183

to the urine-N concentration (9). In addition the true emission seems to
be underestimated by the mass balance method due to a relatively large
background correction. For a reliable prediction of the ammonia emission
more data on the urine-N concentration during grazing at different levels
of N supply have to be collected.
It can be concluded that the average ammonia-N emission from grazed
pastures is limited to a maximum of 10 % of the N returned to the field.
In other words at least 90 % of the excreted N is not lost as ammonia.

(1) VERTREGT, N. and RUTGERS, B. (1988). Ammonia volatilization from
     urine patches in grassland. In: V.C. Nielsen et al. (eds.) Volatile
     emissions from livestock farming and sewage operations pp. 85-91.
     Elsevier Applied Science, London.
(2) MOLEN, J. van der, et al. (1989). Ammonia volatilization from arable
     and grassland soils. In: Hansen, J.AA. and Henriksen, K. (eds.)
     Nitrogen in organic wastes applied to soils. pp. 185-201. Academic
     Press, London.
(3) LOCKYER, D.R. (1984). A system for the measurement in the field of
     losses of ammonia through volatilization. J. Sci. Fd. Agric. 35,
     837-848.
(4) DOAK, B.W. (1952). Some chemical changes in the nitrogenous
     constituents of urine when voided on pasture. J. Agric. Sci. 42,
     162-171.
(5) SHERLOCK, R.R. and GOH, K.M. (1985). Dynamics of ammonia
     volatilization from simulated urine patches and aqueous urea applied
     to pasture. II Theoretical derivation of a simplified model.
     Fertilizer Research 6, 3-22.
(6) VERTREGT, N. and RUTGERS, B. (1988). Ammonia volatilization from
     grazed pastures. CABO report 84, Dutch priority programme on
     acidification, Report 64-2. CABO, Wageningen, 37 pp.
(7) MEER, H.G. van der and UUM-van LOHUIJZEN, M.G. van (1986). The
     relationship between inputs and outputs of nitrogen in intensive
     grassland systems. Developments in Plant and Soil Sciences, 23, 1-18.
     Martinus Nijhoff Publishers, Dordrecht.
(8) MEER, H.G. van der and MEEUWISSEN, P.C. (1989). Emissie van stikstof
     uit landbouwgronden in relatie tot bemesting en bedrijfsvoering.
     Landschap 6, 19-23.
(9) LANTINGA, E.A. et al. (1987). Distribution of excreted nitrogen by
     grazing cattle and its effect on sward quality, herbage production
     and utilization. In: H.G. van der Meer et al. (eds.) Animal manure on
     grassland and fodder crops. Martin Nijhoff Publishers, Dordrecht.
(10) JARVIS, D.C., HATCH, D.J. and ROBERTS, D.H. (1989). The effect of
     grassland management on nitrogen losses from grazed swards through
     ammonia volatilization; The relationship to excretal N-returns from
     cattle. J. Agric. Sci. Camb. 112, 205-216.

GRAZED PASTURES AS SOURCES OF AMMONIA

S. C. JARVIS

AFRC Institute of Grassland and Environmental Research,
Hurley, Maidenhead, Berks., SL6 5LR, UK

Summary
    Details of the extent of NH_3 volatilization from grazed pastures and
the effects of different animal management systems in temperature
grassland are described. Micrometeorological methods have been used to
determine daily, seasonal and annual losses of NH_3 from pastures grazed
by cattle or by sheep and with a wide range of N inputs from both
fertilizers and biological fixation. Wide ranges in the rates of loss were
found over both the shorter and the longer term. Although the variation
in dietary N contents (through an impact on the quantity and form of
excretal N) could be used to explain some of the effects on seasonal NH_3
loss, the mechanisms regulating the changes in daily volatilization rates
were far from clear. The extent of annual losses from grazed pastures
represented only small proportions of N input, and there was a strong
relationship between input and the annual NH_3 volatilization loss. The
results indicated that animal production systems on grazed pastures acted
as sources of NH_3 for at least half the year, and that the patterns of
release were of importance in influencing environmental impact.

1. INTRODUCTION

    The importance of animal production systems as sources of ammonia (NH_3)
has been highlighted in a number of recent reviews and surveys (1,2,3). Whilst
much of this emission relates to the activities and products of housed animals, the
losses associated with animals grazing in the field are thought to be extensive (3).
In principle, it is relatively easy to predict the physical and chemical conditions
conducive to the processes contributing to net NH_3 volatilization when simple
systems are considered. However, once the complexity of the numerous biological
interactions that occur between components of the soil/plant/animal continuum are
imposed upon the physical and chemical equilibria in the equation below,

$$NH_4^+ \rightleftharpoons NH_4^+ \rightleftharpoons NH_3^0 \rightleftharpoons NH_3^0 \rightleftharpoons NH_3^0$$

adsorbed	dissolved	dissolved	gaseous	gaseous
(soil solid	(soil	(soil	(soil	(atmospheric)
phase)	solution)	solution)	atmosphere)	

prediction of losses becomes much more difficult (see 4).
    The major source of NH_3 in grazed systems is from the nitrogen (N) excreted
in dung and urine, particularly that derived from the enzymic hydrolysis of the
urea contained in urine. Other potential sources of NH_3 also exist in both grazed
and ungrazed systems. Plants can release and take up NH_3 *via* their stomata

depending upon environmental conditions and the N status of the plant. Any measured release is the net effect of volatilization and possible absorption by an actively growing sward. Senescing foliage can also release NH_3 and NH_3-based compounds: cut grass left to dry for a few days prior to silage or hay making, and which was not subjected to decomposition, released little or no NH_3, but in humid conditions volatilization amounted to 20-47% of the herbage N content (5). Another potential source is that which occurs directly from applied fertilizer. Losses from ammonium nitrate and ammonium sulphate are relatively small but, depending upon environmental conditions, losses from urea can be substantial, i.e. in excess of 20% of the N broadcast as granules on swards at rates of 50-100 kg N ha^{-1} (6). However, only small proportions of the fertilizer products applied in the UK are as urea; thus only 3.2% of all fertilizer products applied to grassland in 1988 was as urea (7). Volatilization losses from fertilizers (both urea and NH_4^+) were increased substantially when, in controlled environment studies, these were added in combination with urine. Thus, considerably more NH_3 was lost from combined fertilizer and urine additions than was expected from the sum of two components when applied separately (8). This is a further indication of the complexity of the interactions that may occur under normal grassland managements. In the present paper some of the recent research conducted at IGER, Hurley on NH_3 losses from grazed swards is summarised and the implications of the findings discussed.

2. EXPERIMENTAL SYSTEMS

Full details of the methods used to measure NH_3 loss are given elsewhere (4,9). These were based on a mass balance micrometeorological method in which profiles of NH_3 concentrations and wind speeds above the experimental areas were used to calculate daily NH_3 volatilization rates: these were summated to provide period and total grazing losses. The experimental systems were at Hurley where the swards were on a freely drained loam soil overlying chalk at $c.$ 0.7 - 1.0 m. Two experiments were examined. The first was based on a long-term grazing trial in which different swards (0.12 ha) under defined N managements (see Tables 2 and 3) were rotationally grazed for 7-day periods (followed by 21 days recovery) by yearling Friesian steers; for one of the N managements there was also a continuous grazing treatment. In the second experiment, grazing was continuous stocking with ewes to maintain optimal sward height of 6 cm (10) on four 0.65 ha rectangular plots of different sward composition or N management (see Tables 2 and 3). These latter swards were irrigated to minimise the soil water deficit during the grazing season.

3. FACTORS CONTROLLING LOSSES
Environmental changes

The importance of NH_3 volatilization in (i) representing a substantial net loss from production systems, (ii) influencing the oxidation of SO_2 in clouds (11) and (iii) having a major impact on the atmospheric fate of other nitrogenous species (12) underlines the need to understand the surface mechanisms controlling atmospheric NH_3 concentration. Measurements of daily NH_3 fluxes over complete grazing periods provided evidence of a very wide range in rates of loss over both the short (daily) and the longer term (4). Although an attempt at establishing relationships between losses of NH_3 and changes in several environmental factors (wind speed, soil and air temperature, soil moisture status, rainfall and potential evapotranspiration) has been made, the reasons for the daily variability were not clear. Of the measured variables, windspeed had the largest effect, but the best multiple linear regression model had an adjusted R^2 value of only 0.406. The mechanisms regulating the changes in daily losses are therefore far from clear.

Diurnal fluctuations
    Superimposed upon the daily changes in rates of loss were strong diurnal rhythms (13), and concentration gradients above the sward, and therefore the consequent losses, were strongest during the period which included the three hours either side of mid-day (Table 1). This pattern of loss (determined immediately after a grazing period with cattle on ryegrass with 420 kg fertilizer N) was much

	Period	NH_3-N loss kg N ha^{-1}	Proportion of daily total, %
Table 1: Mean (over	15.00 - 21.00 h	0.0115	7
3 days) 6-hourly period	21.00 - 03.00 h	0.0136	8
losses of NH_3 from a	03.00 - 09.00 h	0.0253	15
grazed ryegrass sward	09.00 - 15.00 h	0.1140	70

more distinct than was found when slurry was applied to grass swards (14), and has important implications in relation to mixing processes and chemical interactions in the atmosphere. The same study also provided evidence, in a comparison of two procedures to collect NH_3, which indicated that aerosol loading (i.e. of NH_4^+ in water droplets) was a significant component of normal measurement procedures which should also be considered in routine measurements of loss (13).

Seasonal changes in NH_3 loss
    As well as the shorter-term effects just described, there were major differences in overall extent of NH_3 loss for different periods through the grazing year (4). In early and late season, the trends tended to parallel the grazing pattern and large proportions of the loss occurred whilst the animals were present; this was not the case during other grazing periods. When losses were summed for each period, and results across a number of fertilizer management treatments were considered, there was a strong, positive correlation with the N content of the diet at that time. For 34 sets of observations, the summed losses per grazing period (y; kg NH_3-N ha^{-1}) were related to N content of the herbage (x; %N in dry matter) by the expression:

$$y = -0.0202 + 0.0091 \, x \quad (r = 0.734, P < 0.001)$$

Excretal N returns

The major source for NH_3 loss is provided through the hydrolysis of urea returned in urine and it is of some importance to appreciate the likely extent of this return. Data in Table 2 show the magnitude of the returns and the effects of increasing N input. The total quantity of N recycling through the grazing animals was considerable at all levels of N input to the sward, but increased

Table 2: Annual excretal N returns to pasture from grazing animals

Nitrogen management of sward		N returns, kg N ha⁻¹	
		Total	Urine
Cattle[*]	: grass + 420 kg N ha⁻¹	321	237
	: " + 210 " " "	155	93
	: grass/clover 0N	132	74
Sheep[**]	: grass + 420 kg N/ha⁻¹	425	-
	: grass/clover 0N	255	-

[*](4), [**](10)

significantly as N input to the sward increased. It is also important to note that the proportion of the total N excreted in urine increased in the case of the cattle from 56% in the grass/clover system to 74% with highly fertilized grass. Furthermore, not only did the total quantity of urinary N increase but so also did the urea content as the N content of the diet increased (4). Laboratory scale experiments indicated that changes in other urinary constituents (e.g. hippuric acid) also had a major effect on the extent and pattern of NH_3 volatilization (15).

4. TOTAL LOSSES UNDER GRAZING

Effects of N management

Previous estimates of NH_3 losses suggested that the equivalent of up to 17% of the N input to grazed systems was lost as NH_3 (16). As shown by the data in Table 3, our recent information indicates that the total annual amounts lost during grazing, even at the high animal production rates resulting from high levels of fertilizer input, were relatively small. The annual loss represented, at most, the equivalent of 6% of the N input to the system at high fertilizer rates and this proportion fell as the N input decreased. For the measurements made in the cattle grazing experiment, it is obvious that the extent of overall loss was very much dependent upon the N input to the system. In part, this reflected the increased animal carrying capacity and therefore an increase in the animal returns. However, for the reasons already discussed in relation to the effects of dietary N on the form and distribution of N in excreta, even when the data were expressed on a per animal basis, there were still marked falls in amounts lost as the inputs to the system were reduced.

Table 3. Annual NH₃-N losses from grazed swards

Management	Total loss, kg N ha⁻¹yr⁻¹	Losses per animal, kg N d⁻¹	Proportion of input, %
Cattle:rotational grazing[*]			
: grass + 420 kg N ha⁻¹	25.1	0.018	6.0
: grass + 210 kg N ha⁻¹	9.5	0.009	4.5
: grass/white clover (0N)	6.7	0.005	4.2
Cattle:continuous grazing[*]			
: grass + 420 kg N ha⁻¹	16.3	0.011	3.9
Sheep:continuous grazing[**]			
: grass + 420 kg N ha⁻¹	9.4	0.0012	2.2
: grass + 0N	4.0	0.0010	-
: grass/white clover (0N)	1.1	0.0002	-
: white clover (monoculture, 0N)	11.2	0.0020	-

[*] (9) [**] (17) preliminary data

When data from the cattle experiments at the Hurley site were examined with those from another experiment (with dairy cows) in Lelystad in The Netherlands, the relationship between N loss through volatilization and N input was even more firmly demonstrated (18). In fact, the relationship derived in this study was described by the expression:

$$y = 0.000347 \, x^{1.854} \quad (r = 0.81)$$

where y is the annual loss (kg NH₃-N ha⁻¹) and x is the N input to the system (kg N ha⁻¹).

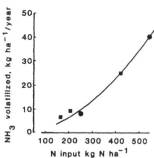

Fig. 1. Relationship between N input and NH₃-N loss at Hurley (■) and Lelystad (●) [Figure from Jarvis and Bussink (18)]

Effects of animal management

It is clear from the preliminary data for sheep grazing systems (Table 3) that the extent of losses were less than with cattle grazing systems with comparable inputs. A number of factors may contribute to this: firstly the urinary excretion patterns (forms and rates) of the two species may differ and this may have had an impact on the potential for loss. Secondly, grazing was continuous with the sheep and, as illustrated by the data in Table 3, in the continuous cattle system losses per animal were less than 70% of those with rotational grazing with comparable N input (although this, in part, may also reflect a decreased ability to monitor losses at low stocking rates). The reasons for this reduction were not clear but may relate to the patterns of grazing in the two managements, their affect on sward heights and the age structure of the sward and therefore the potential for re-absorption of NH_3 by the herbage. Alternatively, the presence of a larger number of animals over a shorter period may have promoted loss because of a greater physico-chemical disturbance, or, because of the greater intensity of excretion, increased the pressure on the various chemical equilibria towards volatilization.

Fertilizer v. fixed nitrogen

The use of clover based swards has been seen as an attractive option to minimise the environmental impact of N leakage from grassland systems. Losses from mixed swards are invariably less than those from highly (i.e. 300-400 kg N ha^{-1} per year) fertilized grass swards because of the lower N input (16). The data in Fig. 1 include that for the mixed grass/clover sward grazed by cattle and demonstrate that it was total N input to the system that was of importance and not whether the N had been biologically or chemically fixed. When inputs from clover were increased, as in the clover monoculture, there were marked increases in losses both on a per area, and, notably, per animal basis (Table 3).

5. CONCLUSIONS

The patterns of release and behaviour of NH_3 from grazed grassland has important consequences for the processes controlling the atmospheric transport and deposition of S and N species and the consequent impact that changes in these components have on environmental quality. The studies at Hurley indicate that for at least half the year, grazed grassland acts as a source of NH_3 and the magnitude of loss increases with increasing N input. Although the extent of this is less than previously thought (16), the patterns of release are of importance in relation to the mixing processes and potential chemical reactions that might occur in the atmosphere. The results also demonstrate (i) that grazed systems with high inputs of N from chemically or biologically fixed nitrogen are equally vulnerable to volatilization losses and (ii) that the mechanisms controlling the extent of net volatilization are complex and poorly understood. It should also be noted that, whilst animals are not grazing and indeed on occasion during grazing on relatively low input systems, opportunity exists for swards to act as sinks for, rather than sources of, NH_3. Further, it is of interest to note that recent calculations (14) have shown that NH_3 losses associated with the housed component of animal production systems (i.e. occurring directly from housing and from stored and applied slurry) can be from 3.5 to 11.4 times greater per animal than during grazing. Whilst the means exist to reduce NH_3 loss from slurry for example (14), there is little current practical potential for reducing grazing losses other than by reducing N intake by the animals and minimising urinary N returns.

REFERENCES
(1) ApSIMON, H. M., CRUISE, M. and BELL, J. N. B. (1987). Ammonia emissions and their role in acid deposition. *Atmospheric Environment*, **21**, 1939-1946.
(2) BUIJSMAN, E., MAAS, J. F. M. and ASMAN, W. A. H. (1985). *Ammonia Emission in Europe*. Rpt R-85-2. Institute for Meteorology and Oceangraphy, Utrecht, The Netherlands.
(3) RYDEN, J. C., WHITEHEAD, D. C., LOCKYER, D. R., THOMPSON, R. B., SKINNER, J. H. and GARWOOD, E. A. (1987). Ammonia emission from grassland and livestock production systems in the UK. *Environmental Pollution*, **48**, 173-184.
(4) JARVIS, S. C., HATCH, D. J. and LOCKYER, D. R. (1989). Ammonia fluxes from grazed grassland: annual losses from cattle production systems and their relation to nitrogen inputs. *Journal of Agricultural Science*, **113**, 99-108.
(5) WHITEHEAD, D. C., LOCKYER, D. R. and RAISTRICK, N. (1988). The volatilization of ammonia from perennial ryegrass during decomposition, drying and senescence. *Annals of Botany*, **61**, 567-571.
(6) GOULD, W. D., HARGEDORN, C. and McGREADY, R. G. L. (1986). Urea transformations and fertilizer efficiency in soil. *Advances in Agronomy* **40**, 209-238.
(7) SURVEY OF FERTILISER PRACTICE (1988). Fertiliser Use on Farm Crops in England and Wales, 1988. ADAS, London.
(8) MEYER, R. D. and JARVIS, S. C. (1990). The effects of fertiliser/urine interactions on NH_3-N losses from grassland soils. *Proceedings XVI International Grassland Congress*, Nice 1990, pp. 155-156.
(9) JARVIS, S. C., HATCH, D. J. and ROBERTS, D. H. (1989). The effects of grassland management on nitrogen losses from grazed swards through ammonia volatilization; the relationship to excretal N returns from cattle. *Journal Agricultural Science*, **112**, 205-216.
(10) PARSONS, A., PENNING, P., ORR, R. and JARVIS, S. (1987). Are grass-clover swards the answer to nitrogen pollution? *In:* Science, Agriculture and the Environment (ed. J. E. Y. Hardcastle) pp. 10-11. AFRC, London.
(11) SCHWARTZ, S. E. (1984) Gas aqueous reactions of sulphur and nitrogen oxides in liquid-water clouds. In: SO_2, *NO and* NO_2 *Oxidation Mechanisms* (ed. J. G. Calvert). Butterworths, Boston.
(12) DERWENT, R. G., DOLLARD, G. J. and METCALFE, S. E. (1988). On the nitrogen budget for the United Kingdom and North-West Europe. *Quarterly Journal of the Royal Meteorological Society*, **114**, 1127-52.
(13) HATCH, D. J., JARVIS, S. C. and DOLLARD, G. J. (1990). Measurements of ammonia emission from grazed grassland. *Environmental Pollution* (in press).
(14) JARVIS, S. C., PAIN, B. F., HATCH, D. J. and THOMPSON, R. B. (1989). Ammonia volatilisation and loss from grassland systems. *Proceedings XVI International Grassland Congress, Nice*, 1989, pp. 157-158.

(15) WHITEHEAD, D. C., LOCKYER, D. R. and RAISTRICK, N. (1989). Volatilisation of ammonia from urea applied to soil: influence of hippuric acid and other constituents of livestock urine. *Soil Biology and Biochemistry*, **21**, 803-808.

(16) WHITEHEAD, D. C., GARWOOD, E. A. and RYDEN, J. C. (1988). The efficiency of nitrogen use in relation to grassland productivity. *AGRI Ann. Rpt. 1985-86*, 86-89.

(17) JARVIS, S. C., HATCH, D. J. and ORR, R. J., unpublished, preliminary data.

(18) JARVIS, S. C. and BUSSINK, D. W. (1990). Nitrogen losses from grazed swards by ammonia volatilization. *Proceedings of European Grassland Federation Meeting, Czechoslovakia, 1990* (in press).

Session VI

DEVELOPMENTS IN ODOUR MEASUREMENTS

Chairperson : V.C. NIELSEN

Odour Emissions from Broiler Chickens

C. R. Clarkson[1,2] and T. H. Misselbrook[1]
[1]AFRC Institute for Grassland and Animal Production,
Hurley, Maidenhead, Berkshire SL6 5LR, UK
[2]Attached to IGAP from ADAS Liaison Unit, Wrest Park,
Silsoe, Bedfordshire MK45 4HS, UK

Summary

Olfactometric techniques were used to assess odour concentration, emission and intensity of samples taken from the exhaust air of commercial broiler houses under different management regimes. Results show that odour concentration and emission increased sharply after the crop had reached 30 days of age. Low moisture content litters produced lower odour emissions than high moisture content litter. A novel heating system that enables greater ventilation rates than normal may reduce odour emission to acceptable levels. Intensity measurements indicate that the exhaust air odour concentration must be reduced by > 97% to reduce its intensity category to 'faint odour'. This may be difficult to achieve with abatement devices only.

1. Introduction

Odour emissions from UK agricultural enterprises are the cause of many complaints from the public to Environmental Health Officers. Poultry units account for nearly 25% of all complaints, with approximately 300 premises causing 900 complaints per year. The majority of these premises are broiler units which cause odour nuisance during the last few weeks of the production cycle. Previous workers[1,2] have shown that odour and ammonia emissions rise steeply once the birds are 3-4 weeks of age. This rapid increase in emission may be due to the formation of a dry, impermeable 'cap' on the litter which results in a reduction in moisture absorbing capacity[3]. Theoretically it may be possible to introduce fresh wood-shavings at this point or to break up the capped layer of litter in an attempt to increase its moisture carrying capacity, but practically this is not easy to accomplish.

To assess the likelihood of an odour source causing a nuisance, quantitative information about the odour stream is required. Quantitatively, odour emission rates are important for an understanding of the atmospheric transport of the odour plume, and can be used in dispersion models to predict downwind odour concentrations[4]. Odour intensity is a measure of the subjective strength of an odour and can be used to assess the degree of abatement required to reduce the likelihood of nuisance.

This paper describes experiments in which samples of exhaust air from broiler houses, operating under different management regimes, were collected and analysed by olfactometry for odour concentration, emission and intensity.

2. Materials and Methods

2.1. Sites.

Odour samples were taken from 3 different sites between 1987 and 1989. Details of each site are given below.

Site A was located in N. Wales. Samples were taken in 1987 and 1988 to compare odour emission between houses fitted with traditional 'bell' drinkers or 'cup' type drinkers. Apart from the drinker type, other conditions were identical. Each building housed approximately 30,000 birds, the ventilation system comprising roof-mounted extracting fans, and the heating was *via* gas brooders.

Site B was located in Northamptonshire. Samples were taken in early 1989 throughout one crop to compare odour emissions with crop growth. The house as filled with a computerised ventilation system *via* roof-mounted extracting fans controlled by humidity and temperature sensors within the house. The house contained 40,000 birds at the start of the crop, and was heated indirectly by oil-fired space heaters.

Site C was located in Gloucestershire. Samples were taken in autumn 1989 to compare odour emissions with crop development. This site employed a novel heating system where composted broiler litter was burned to produce steam. The steam was used to heat broiler houses *via* a piped heating system[5]. As the heat costs very little, higher ventilation rates than normal can be used to lower relative humidities within the broiler house. The experiment was conducted to see if this system had any effect upon odour emission. This broiler house was fitted with roof-mounted extracting fans and housed 26,000 chickens before thinning at day 44.

Site D relates to a site investigated by the Warren Spring Laboratory in autumn 1980 and is included for purposes of comparison. The house was heated by gas brooders, was ventilated by roof-mounted extracting fans and contained 9,000 broiler chickens.

2.2. Odour measurements

Odour samples were collected from sites A, B and C from the ventilation exhaust air by evacuating a rigid air-tight vessel containing an empty 60 l Teflon bag, whose inlet was connected *via* a PTFE sampling line to the odour stream. Odour measurements were made with two dynamic dilution olfactometers (Olfaktomat - Project Research, Amsterdam). The presentation method was a forced choice from two odour cups to a panel of normally 8 people. All procedures and equipment conformed to the current EEC olfactometric guidelines[6].

2.3. Odour concentration

The 50% odour threshold dilution was determined for each sample within 24 hours of collection using procedures described by Pain *et al.*,[7].

2.4. Odour intensity

A direct scaling technique was employed using the categories described by Paduch[8]. The scale used by the odour panellists to describe the subjective strength of a sample was as follows:

0	No odour
1	Very faint odour
2	Faint odour
3	Distinct odour
4	Strong odour
5	Very strong odour
6	Extremely strong odour

The procedure entailed presenting up to 10 supra-threshold diluted air samples to the panel in random order, inter-spaced with blanks. The 75% detection threshold concentration was used as the least concentrated sample. One odour port of the olfactometer passed clean 'reference air' at all times. The process was repeated 3 times for each measurement. To calculate the result, the mean intensity score for each concentration step was regressed against the logarithm of the relative concentration - Weber-Fechner Law[9].

2.5. Odour emission

The ventilation rate was measured at each site whilst the odour samples were being collected. Odour emission is the product of the odour concentration of the air and its flow rate ($E = DF$, where E = odour emission OU s^{-1}, D = 50% dilution threshold OU m^3 and F is the odour stream volumetric flow rate m^3 s^{-1}).

3.   Results and Discussion

3.1.   Odour concentration from broiler houses fitted with different drinker types - Site A

On both sampling occasions, the odour threshold concentration of exhaust air from the house fitted with mini-cup drinkers was substantially less than that from the house equipped with conventional 'bell' drinkers - see Fig. 1. The rationale for fitting mini-cup drinkers was that it was thought that the mini-cup drinkers would reduce water wastage and hence improve the litter conditions. A visual inspection of both houses confirmed that dryer, more friable litter was produced in the mini-cup equipped house on both occasions, although no data on litter moisture content is available. Whilst it is not possible to state that the drinker type is solely responsible for the difference in odour concentration from just two measurements, it seems possible that drinker type (and hence litter conditions) may be an important factor.

3.2. Odour concentration and emission throughout broiler crop - Sites A, B, C

Figures 2 and 3 show that the odour concentration of the exhausted air and odour emission generally increase as the birds age. Anecdotal evidence from poultry specialists concerning complaints about odour nuisance points to the concentration and emission patterns of Site B as being 'typical' for a UK broiler unit, where sharp increases in concentration and emission occur when the crop is around 30 days old. Site C is atypical in this respect, in that odour concentration and emission are low and fairly constant throughout the crop. Data from Site D has been included for comparative purposes. The variation in odour concentration recorded on this site after day 30 is partially explained by large variations in the ventilation rate and ambient weather conditions encountered when the samples were taken. Nevertheless, the odour emission shows a sharp increase after day 43.

The litter moisture content was measured throughout the crop on Sites B and C. Site B is considered typical with a steady rise throughout the crop with the ultimate moisture content exceeding 50% (Fig. 4). Site C is atypical as the moisture content does not exceed 30%. As stated above in 3.1, the dry friable litter is associated with low odour concentration and emission (Figs 2 and 3). It is thought that the enhanced heat input and ventilation rates associated with this novel heating system were directly responsible for the low moisture content of the broiler litter, odour concentration and odour emission.

Ammonia emission was assessed at site C and the results are presented in Fig. 5. Data from van Ouwerkerk and Voemans[1] are presented for comparison purposes. Compared with the odour emission from Site C, the NH$_3$ emission does not seem to

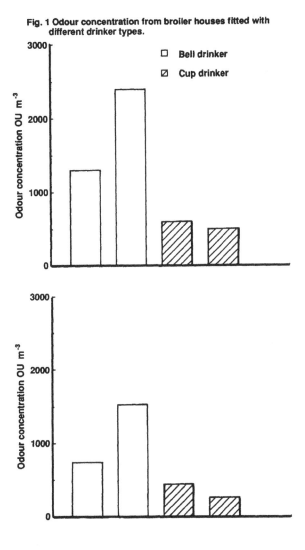

Fig. 1 Odour concentration from broiler houses fitted with different drinker types.

Fig. 2 Odour concentration of exhaust air throughout broiler crop

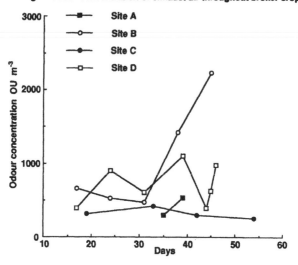

Fig. 3   Odour emission throughout boiler crop

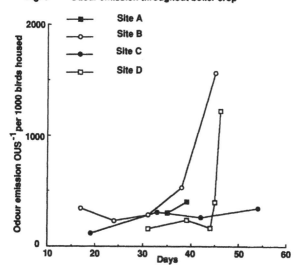

Fig. 4 Litter moisture content throughout broiler crop

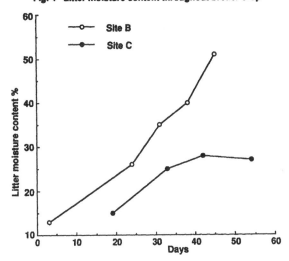

Fig. 5 Ammonia emission throughout broiler crop

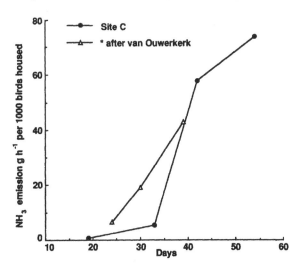

have been inhibited by the dry litter conditions, hence the mechanisms that drive odour and ammonia emission may be independent of each other.

3.3. Odour intensity - Sites B and C

The intensity *vs* concentration relationship is of the form $I = a \log_{10}$ Rel. Conc. + b, where I is the intensity category (as assessed by the Odour Panel), a is the slope of the regression line, Relative Concentration is $1/D_{50}$ and b is a dimensionless constant. This relationship is shown graphically in Fig. 6. Values of a, b and r^2 are given in Tables 1 and 2 for sites B and C respectively.

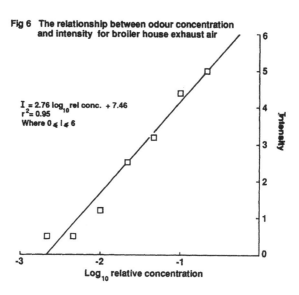

Fig 6 The relationship between odour concentration and intensity for broiler house exhaust air

$I = 2.76 \log_{10}$ rel conc. $+ 7.46$
$r^2 = 0.95$
Where $0 \leqslant I \leqslant 6$

Log_{10} relative concentration

Table 1: Odour Intensity - Site B

Day no.	a	b	r^2	Reduction in odour concentration to reduce intensity to faint odour %
10a	2.7	7.2	.92	98.9
17a	2.2	6.3	.94	98.8
24	2.8	7.5	.95	98.9
31	2.7	7.8	.97	99.3
38	2.6	7.7	.93	99.4
45	2.5	7.5	.95	99.3

Note: (a) Days 10 & 17 significantly different from days 24-25 at 95% level

Table 2: Odour Intensity - Site C

Day no.	a	b	r^2	Reduction in odour concentration to reduce intensity to faint odour %
19	1.8	5.8	.89	99.2
33	2.4	6.0	.93	97.8
42b	2.7	6.6	.91	98.0
54b	2.5	6.8	.90	98.8

Note: (b) Day 42 and 54 significantly different from days 19 & 33 at 95% level

Also included in the above Tables is the percentage reduction in odour concentration (abatement) in threshold concentration of the exhaust air required to achieve an odour intensity of 2 (faint odour). It is thought that this may be an acceptable level in the assessment of an odour as a nuisance. In the above experiments, at least 97% abatement is required to reduce the intensity to faint odour. Abatement can be achieved by atmospheric mixing of the odorous steam with ambient air or by an abatement device such as a bio-scrubber of filter. In either case, this type of intensity data is especially important in determining minimum distances between the source and a potential complainant with the aid of dispersion modelling.

Given the present 'state of the art' in odour abatement technology, it is unlikely that any single device could operate at such high (> 97%) odour abatement efficiencies. Therefore a combination of atmospheric mixing and abatement devices may be necessary to achieve a sufficient odour concentration reduction.

Conclusions
The following conclusions can be drawn from the above experiments:
1. Odour concentration and emission increase throughout a broiler crop's life, rising steeply after 30 days.
2. Litter moisture content is an important factor in odour emission, but ammonia emission is primarily driven by other factors.
3. Novel heating systems that allow increases in ventilation rate and reduce litter moisture contents may reduce odour emissions to acceptable levels.
4. Odour intensity measurements provide a practical method for the determination of the degree of abatement needed to reduce odour nuisance.
5. The mean reduction in odour concentration required to reduce broiler house exhaust air to a 'faint odour' was 98.8%. This may be difficult to achieve with abatement devices only.

References

1. Van Ouwerkirk, E. N. J. and Voemans, J. A. M. (1986) The effect of insulating broiler house flows on odour emission. *In:* Nielsen, V. c., Voorburg, J. H. and L'Hermite, P., eds. 'Odour prevention and control of organic sludge and livestock farming'. Elsevier Applied Science Publishers, London. pp 175-180.

2. Kroodsma, W., Scholtens, R. and Huis in 't Veld, J. (1988) Ammonia emission from poultry housing systems. *In:* Nielsen, V. C., Voorburg, J. H. and L'Hermite, P., eds. 'Volatile emissions from livestock farming and sewage operations. Elsevier Applied science Publishers, London. pp 152-161.

3. Charles, D. (1990) MAFF Nottingham U.K. Personal Communication.

4. Williams, M. L. and Thompson, N. (1986) The effects of weather on odour dispersion from livestock buildings and from fields. *In:* Nielsen, V. C., Voorburg, J. H. and L'Hermite, P., eds. 'Odour prevention and control of organic sludge and livestock farming'. Elsevier Applied Science Publishers. London. pp 227-233.

5. Dagnall, S. and Mann, R. (1989) ADAS Poultry Journal 3. pp 46-58.

6. Hangartner, M., Hartung, J., Paduch, M., Pain, B. F. and Voorburg, J. H. (1989) Improved recommendations on olfactometric measurements. Environmental Technology Letters 10, 231-236.

7. Pain, B. F., Phillips, V. R., Clarkson, C. R., Misselbrook, T. H., Rees, Y. J. and Farrant, J. W. (1990) Odour and ammonia emissions following the spreading of aerobically treated pig slurry on grassland (in press).

8. Paduch, M. (1988) Present state of VDI guidelines on odour assessment. *In:* Nielsen, V. C., Voorburg, J. H. and L'Hermite, P., eds. 'Volatile emissions from livestock farming and sewage operations'. Elsevier Applied Science, London. pp 38-53.

9. Hangartner, M. (1989) Scaling of odour intensity. *In:* Proceedings of Workshop of the COST 681 Odours Group. Zurich April 1988, CEC. pp 77-85.

FIELD MEASUREMENTS OF ODOROUS AIR POLLUTION WITH PANELS

Hendrik Harssema

Dept. of Air Pollution
Wageningen University
POB 8129
NL-6700 EV Wageningen
The Netherlands

Summary

A sub-group of the COST 681 Working Group on Olfactometric Measurements was asked to harmonize odour measurements with panels in ambient air. In this paper an inventory is presented of methods which are used in various European countries. This inventory is followed by a critical discussion of the differences between measuring strategies, selection and composition of the panels and of the set-up of measurements. A number of recommendations concerning the execution of field measurements of odorous air pollution with panels are presented together with a glossary of terms to be used by technicians and psychologists in the field of odour measurements.

1. PREFACE

The COST 681 Working Group on "Olfactometric Measurements" decided in October 1988, that an ad-hoc working group should be initiated in order to harmonize measurements concerning "the approaching of sources in the field with a panel". The Working Group considered this "...... as a very practical estimate of the odour nuisance. It is worthwhile to also harmonize these techniques".

The ad-hoc working group first redefined the subject somewhat, in order to keep the field of interest within limits. The group then collected information on the existing techniques. This information is described in this paper. At the Silsoe meeting in March 1990 consensus was reached on a number of recommendations.

The ad-hoc group consisted of the following persons:

Birgitta Berglund (Sweden)
Hendrik Harssema (Netherlands, chairman)
Eckehard Koch (Fed. Rep. Germany)
Heinrich Mannebeck (Fed. Rep. Germany)
Marie-Line Perrin (France)
Stephen Toogood (United Kingdom)

During the meeting of this ad-hoc group, the difference in language between technicians and psychologists arose as an important topic. The simple word "odour" appeared to be defined in different ways. Technicians use this word quite loosely whenever they mean something that smells, whereas psychologists strictly reserve this word for the sensation of smelling. In order to avoid misunderstandings, a glossary of terms, based on a proposal by Birgitta Berglund is included at the end of this paper.

2. INTRODUCTION

At first sight odour observations in the field seem to be a very direct way to establish the magnitude of the problem area around a source of odorous substances. At large distances from the source no odour will be detectable, so by approaching the source in the field the distance at which this source is perceptible can be determined. Unfortunately, this is not so simple as it looks, and therefore a more extensive analysis of the subject should underpin a description of available methods.

In a discussion on odour problems it is useful to make a strict distinction between **emission**, **exposure**, **detection** and **annoyance** (1). Although these aspects form a causal chain, their relation is quite complex. Without odour emission there can be no odour annoyance, but the presence of odour emission does not necessarily lead to odour annoyance. To make an odour problem manageable, a proper quantification of the various steps in the causal chain is needed.

Odour emission can be measured by olfactometric methods (2). The most simple situation is that of a well defined airstream containing odorants. In a sample from this stream the odour concentration can be measured. Multiplication of this concentration with the flow rate gives the odour emission rate. Quantification of odour emissions from non-point sources is less straightforward; in that case field measurements may be needed instead.

Odour exposure is a physical quantity which can be determined by **calculation** or by **measurement**. Calculation is done with a dispersion model with the emission rate as an input, together with meteorological data and information on the source and the terrain. Dispersion models for odorants may be based on a fluctuating Gaussian plume model (3) and (4) or on assumptions on the frequency distribution of instantaneous concentrations (5). The result of such calculations is the frequency of odour concentrations above threshold. By using the statistics of meteorology, also the statistics of this odour concentration can be determined. Direct immision measurement is another way of quantifying odour exposure. Since this is the subject of this paper, it will be amplified in the following sections.

Odour detection is a subjective quantity. It deals with human receptors who, in their living environments, become aware of the presence of odorants. Odour detection is a momentary yes-or-no response and occurs when the odour concentration exceeds the detection threshold. The odour is recognized at the recognition threshold, which is generally at a higher concentration than the detection threshold.

Odour perception is a subjective construct which refers to various aspects of the supra-threshold response of the human olfactory system. When an odour is detected in inhaled air it is possible to judge its percieved intensity or how strong it smells. It is also possible to recognize the odour by its percieved quality, for example by classifying the odour as being the smell of manure or the smell of rotten eggs. The odour intensity is measured by judgemental methods among which the method of category scaling is the most used one in field studies.

It is **odour annoyance** that is to be eliminated in odour control schemes. It may be the result of a history of repeated detections of odour but also of the percieved intensity or percieved quality of the odorous air inhaled. Generally, odour annoyance is considered to be the result of the human evaluation of the odour exposure rather than an odour perception as such. Apart from the time history of detections, a great number of other variables play a role in the relation between exposure, perception and annoyance. These variables may originate from the olfactory system, the personal make-up and the social setting. It should be stressed, that odour annoyance can only be adequately measured in a population exposed to odorous air pollution in their real-life situation.

Measurement of exposure is the subject of this report. As shown above, it essentially means the determination of odour concentration in ambient air. In analogy to the measurement of emission, standard olfactometric methods may be applied by taking samples and analysing them in the odour laboratory. By using a mobile laboratory these measurements can even be performed "in the field" (see e.g. (6)). The generally low and variable concentrations in the field may cause specific problems with this approach, but the general recommendations for olfactometry also apply to these measurements (2). Therefore, this report will not deal with field measurements performed in mobile laboratories, but be limited to those measurements, where observers make odour measurements in the ambient air.

The subject of this report will be limited to **sensory measurements**, so no annoyance reactions will be dealt with.

3. INVENTORY OF METHODS

Field measurements of odorous air pollution can be **source- or receptor-oriented,** depending on the objectives. Source-oriented measurements are generally performed in order to estimate the impact of the source in the surroundings. Receptor-oriented measurements are aimed at determining the exposure of a community.

Source oriented measurements quite often use a plume concept and try to determine the detectable part of an odorous plume. They are always done downwind of the source under examination. Although there is a wide variety in the way of performing the measurements, **two main groups** can be distinguished. In one approach, the field panel (or sniffing team) is **continuously moving around** downwind the source, trying to determine the outline of the detectable part of the plume. In the second approach, **fixed positions** are sampled for some time (order of minutes), and some statistical measure of the detectability is measured.

An example of the **first approach** is given by Graafland, et.al. (7). At a number of distances from the source, starting at a large distance at which no detection of odour is expected, the panel walks along a line more or less perpendicular to the plume axis. The length of this line is such that the axis of the momentary plume is with certainty included, and each panel member must pass this axis at least once. For each line, each panel member indicates, whether odour was detected at least once, or not at all. If 50 percent or more of the panel has detected odour, the odour is assumed to be present at that distance. The result of one measuring cycle is the "odour distance", the largest distance at which the odour was detected by 50 percent of the panel. Lindvall (8) determined the **momentary width of an odorous plume** by driving with cars through a cross-section of the plume. The observers were seated in the cars with the windows open.

The **second approach** was first described by Grennfelt and Lindvall (9). A number of observers are for some time located at fixed positions on a line, perpendicular to the plume axis. They make a number of momentary observations on the presence and the intensity of the odour. Intensity is scaled by using magnitude estimation with a standard reference. Variations of this method were used by Thiele, et. al. (10), Medrow (11) and DeBree and Harssema (12). The methods are different in the number of observers, the time of one measuring cycle, the frequency of observation and the method of intensity scaling. All methods, however, yield an **"odour frequency"** at each measuring location; this odour frequency is defined as the percentage of the observation time that odour was detectable.

Receptor oriented measurements may be performed irrespective of wind direction or in a district downwind of a complex area source. They are not intended to cover a plume, but to cover an area. Panel members are located at fixed positions during a couple of minutes and evaluate the presence and sometimes also the intensity and quality of odour. Examples of this approach have been presented by Perrin, et.al. (13) and Winneke and Paduch (14). Population panels as described by Köster, et.al. (15) also determine the presence of odorants in the field, although the main objective of these measurements is the determination of annoyance.

Both approaches are characterized by a large variety in the way of performing the measurements. Each research group has developed its own method and has often quite good reasons for the choices made. In the Federal Republic of Germany an attempt is being made to standardize field measurements of odorous air pollution in a VDI Working Group (16). In the draft recommendation both approaches mentioned above are described. Koch & Schütz (17) provide a discussion (in German) of the proposed methods. The French AFNOR committee on olfactometry, in an attempt to standardize intensity measurements, also prepares recommendations on field measurements (Perrin, personal communication, 1990).

In the next section the main differences between the methods will be discussed.

4. DISCUSSION OF METHODS

Methods for field measurements of odorous air pollution vary in the **measuring strategy, the selection and composition of the panel** and the **set-up of a measurement**. Therefore the discussion in this section will be along these lines.

Measuring strategy

The measuring strategy determines **when** and **where** the measurements are performed. As shown in the previous section, two main points of variation exist: fixed versus variable locations and plume oriented versus area oriented locations.

Performing the measurements at **fixed locations** should generally be preferred, since the odour concentration in ambient situations is often quite variable. When moving around in a concentration field which is variable in time and space, the results are difficult to interprete in a spatial framework. This is especially the case downwind of a point source, where variations in concentration are more pronounced than downwind of an area source. When the measurement is aimed at getting a representative result for an area, random movement within the area may be useful, but even in that case observations should be taken at well defined timesteps and locations.

The **duration** of a measurement is closely linked to the measuring strategy, but will be discussed under the heading of the measurement itself.

The choice between a **plume oriented** and an **area oriented** approach is completely determined by the objective of the measurement. Plume oriented measurements should cover the whole plume, or at least the time averaged plume axis. Since the actual position of the plume axis is quite variable in practice, simultaneous measurements at a number of locations perpendicular to the plume axis are necessary in order to determine this position and to get information on the plume cross-section. Area oriented measurements should of course cover the area of interest in a representative way. Simultaneous measurements at different locations are not necessary in the latter case, but care should be taken to avoid systematic correlations between a turning wind direction or changing meteorological conditions (day/night cycle) and the location of the measurement.

The panel

The panel should be defined in an absolute way with respect to the sensitivity of the general population, and at the same time it should be as homogeneous as possible, in order to obtain reproducible results. Therefore, some selection of panel members is advisable.

Perrin (personal communication, 1989) makes a selection based on the individual thresholds of perception for five pure compounds; panelists with a threshold which is more than a factor of 10 smaller or larger than the average for any compound are excluded, in accordance with the French AFNOR standard. Also, they must be able to discriminate between the reference scale intensities.

VDI 3940 (16) gives an extended discussion on panel selection. Panelists should be between 18 and 50 years of age; sex and smoker/non smoker is considered as of no importance. Sensitivity to odours must be examined; individual thresholds for hydrogen sulfide and n-butanol are to be measured and compared to the average. No value is prescribed, however, of the admissible deviation from this average. Also the ability of the panelists to scale the intensity and character of odours, their health and their ability to bear responsibility and to concentrate are mentioned as important selection criteria.

De Bree and Harssema (12) selected their panelists according to their individual threshold for hydrogen sulfide; when the threshold deviated more than a factor of three from the average, they were excluded. In the Dutch preliminary guideline for olfactometry an individual threshold value for n-butanol is used as a criterion; a panelist is excluded when his threshold deviates more than a factor two from the average (18). This same criterion is going to be applied to members of sniffing teams.

Since generally field measurements of odorous air pollution are performed with experienced panels, new panel members should be trained in order to avoid inhomogeneity. It is very important to check the performance of the panel at regular intervals, so that its sensitivity is defined in an absolute way at any moment.

Panel size is a compromise between what is theoretically desirable and what can be afforded. Grennfelt and Lindvall (9) used 25 to 28 panelists, Graafland, et.al. (7) suggest a panel size of at least 10. De Bree & Harssema (12) generally use a panel of 8 persons. VDI 3940 (16) assumes the existence of a "pool" of panelists, from which the actual panel is chosen. The size of the panel is allowed to depend on the specific problem for which the measurements are set up. Perrin (personal communication, 1990) generally uses a panel of 6 to 8 persons.

The set-up of the measurements

In general, one measurement consists of a number of observations by one panelist at one location. Such a measurement takes 5 to 15 minutes. Although this duration of one measurement is only seldom discussed, it is generally taken to be representative for a specific meteorological condition. Strictly speaking, the measurement should last for one hour in that case, since that is the basic averaging time for meteorological measurements. Making odour observations for one hour without interruption, however, demands too much of the panelists' attention, so the period of 5 to 15 minutes can be regarded as a compromise.

During each measurement either the duration of detectability of odours is determined or a number of observations at fixed time-intervals is made. With both methods the frequency of odour detectability relative to the observation period is the outcome of one measurement. Depending on the objective of the study, a number of measurements must be performed in order to get a representative set of data.

Taking only **detectability** into account is the most simple form of field measurements of odorous air pollution, but in almost every method used also **odour intensity** is somehow measured and sometimes even **odour quality**.

The latest draft of the VDI-method for the Federal Republic of Germany (16) prescribes at least a yes/no response, but recommends the use of a 7-point intensity scale (0 = no odour, 1 = very weak odour, ..., 5 = very strong, 6 = extremely strong). In Northrhine-Westfalia the latest opinion is, that no intensity or quality measurents should be performed; it is assumed to be sufficient to determine whether odour occurs, yes or no, and at what frequency it occurs (19).

De Bree & Harssema (12) use a 4-point category scale, but in general, they do not measure odour intensity; their use of the intensity scale is primarily to facilitate the choice of the panelists.

Perrin et. al. (13) make intensity measurements by scaling relative to a reference scale consisiting of a number of pyridine-in-water samples (generally 4), which the panelists take with them in the field. From the intensity values reported, pyridine-equivalent concentrations are calculated.

Berglund (personal communication, 1989) also advocates the combination of detectability and intensity measurements (preferrably with reference scales) in the field, since this combination has yielded interesting insights in the laboratory. A main problem is, that for a proper interpretation blanks should also be presented to the panelists, and this is quite difficult in the field. Therefore, the panelists' response to signal-and-blank should be determined in the laboratory.

5. RECOMMENDATIONS

Recommendations for the performance of field measurements of odorous air pollution with panels cannot be as complete and strict as they are for the measurement of odour concentration in the laboratory for two main reasons. First, far less experience is present for the former than for the latter and second, the objectives for field measurements are much more diverse than they are for odour concentration measurements. Nevertheless, the ad-hoc working group on Field Measurements of Odorous Air Pollution with Panels reached an agreement on the following recommendations:

* Field measurements of odorous air pollution with panels are a useful tool in characterizing ambient air quality with respect to odorants. They yield information at the **exposure level** which is essential in linking emission to perception and eventually to annoyance.

* Subjects, who are going to perform field measurements of odorous air pollution should be **selected**. There should not be a large variation in odour sensitivity within the panel. Therefore each subject's individual threshold for preferrably four, but at least one component should be determined at regular intervals, so that their sensitivity is well defined. Subjects with a personal threshold which deviates more than a factor three from the average should be excluded.

* Measurements should be performed in such a way, that panelists are at a fixed location for at least 10 minutes. During this time-interval the fraction of time that odour is detectable should be determined. The most adequate way of obtaining this information is by making at least 20, but preferrably more, observations at fixed time intervals (e.g. every 10 seconds).

* The result of one measurement at one location should be expressed as the **odour frequency**, which is defined as the time during which odour is detected relative to the total observation time.

* A subject should not take measurements for longer than 20 minutes at the most, without taking a rest of at least the same duration.

* Whether a panel should be trained or not depends on the objective of the study. A trained panel works more efficiently and makes the results from different studies more comparable.

Depending on the objective of the measurements, more information can be gathered, for instance on odour intensity and/or odour quality. When **odour intensity** is measured the following **additional recommendations** apply:

* Use of reference scales is preferred over category scales.

* Only those panelists should be selected for intensity measurements in the field, who can discriminate the odour intensities of a set of at least 5 concentrations of a reference substance.

* Results of intensity measurements should be expressed in equivalent concentrations of the reference scale(s) used.

GLOSSARY OF TERMS:

Physical terms:

Substance	Species of matter of definite chemical composition.
Odorant	A substance which stimulates an olfactory system such that an odour is perceived.
Chemical	A substance produced by or used in a chemical process.
Chemical compound	A pure substance composed of two or more elements, with a constant composition.
Odorous chemical compound	A chemical compound which stimulates an olfactory system such that an odour is perceived.
Air	Mixture of gases, with nitrogen and oxygen as the main substances.
Air pollutant	Something that pollutes air, for example a chemical compound that contaminates air.
Air pollution	The presence of air pollutants in the atmosphere.
Odorous air	Air that stimulates an olfactory system such that an odour is percieved.
Odorous air pollution	Air pollutant that stimulates an olfactory system such that an odour is perceived.
Emission	The discharge of pollutants at the pollution source.

Odour emission	The discharge of odorous pollutants at the pollution source.
Immission	The presence of pollutants in the living environment.
Odour immission	The presence of odorous pollutants in the living environment.
Exposure	A quantitative measure of the presence of pollutants in the environment of a subject.
Odour exposure	A quantitative measure of the presence of odorous pollutants in the environment of a subject.
Concentration of air pollutants	Amount of air pollutants per unit volume of air (mass or volume per unit volume).
Odour concentration	Amount of odorants per unit volume of air, expressed in odour units per cubic metre of air.
Odour unit	The amount of (a mixture of) odorants which, after mixing with odorant-free air to a volume of one cubic metre, is just detected as odorous by fifty percent of the members of an odour panel.

Psychological terms:

Sensation	An experience arising directly from stimulation of sense organs.
Odour	A sensation resulting from adequate stimulation of the olfactory system, a smell.
Odour detection	To become aware of the sensation resulting from adequate stimulation of the olfactory system.
Odour detectability	The probability of an odour being detected by an olfactory system.
Odour frequency	The percentage of the observation time that odour detection occurs.
Scent	Distinctive odour, especially when agreeable.
Fragrance	Odour of pleasant quality.
Pleasant odour	Sweet-smelling odour.
Unpleasant odour	Disagreeable odour.
Odour perception	The act of perceiving by the olfactory system. A meaning obtained by an olfactory process while an odorous stimulus is present.
Odour percept	The mental result or product of perceiving by the olfactory system.
Odour intensity	An attribute or fundamental characteristic of an odour percept is its intensity. The perceived odour intensity refers to how strong an odour smells.
Odour quality	The characteristic nature of a particular odour percept; it refers to what it smells like (roses, pulp mill exhaust, manure, etc.).
Annoyance	The feeling of being annoyed: a nuisance. Annoyance can only be evaluated by a human being.
Odour annoyance	The feeling of being annoyed by odours.
Smell	To give out an odour or to use the sense of smell.
Smeller	One who tests by smelling.
Odour panel	Any list or group of persons serving as smellers.
Odour panelist	A member of an odour panel.

REFERENCES
(1) HARSSEMA,H. (1987). Characterization of exposure in odour annoyance situations. In: H.S.Koelega (Ed.), Environmental annoyance, 95-104. Elsevier, Amsterdam, 1987.
(2) VOORBURG,J.H. (1986). Recommendations on olfactometric measurements. In: V.C. Nielsen, et.al. (Eds.), Odour prevention and control of organic sludge and livestock farming, 378-381. Elsevier, Amsterdam.
(3) HÖGSTRÖM,U. (1972). A method for predicting odour frequencies from a point source. Atmosph. Environment 6, 103-121.
(4) DE BREE,F.B.H. & H.HARSSEMA (1988). Field evaluation of a fluctuating plume model for odours with sniffing teams. In: K. Grefen & J. Löbel (Eds.), Environmental Meteorology, 473-486.
(5) MEDROW,W. & C. JüRGENS (1984). Die Simulation der Geruchsausbreitung. Staub Reinhalt. Luft 44, 475-479.
(6) BERGLUND,B., U.BERGLUND & L.LUNDIN (1988). Odour reduction by biological soil filters. In: R.Perry & P.W.Kirk (Eds.), Indoor and ambient air quality, 410-419. Publ. Div. Selper Ltd., London.
(7) GRAAFLAND,T.F., C.J.M.ANZION, & A.J.DRAGT (1987). Bruikbaarheid snuffelploegmetingen bij stankonderzoek. Publikatiereeks Lucht nr. 66. Ministerie VROM, 's Gravenhage.
(8) LINDVALL,Th. (1974). Monitoring odorous air pollution in the field with human observers. Annals New York Acad. of Sciences 237, 247-260.
(9) GRENNFELT,P. & LINDVALL,T. (1973). Sensory and physical-chemical studies of pulp mill odors. Proceedings Third Clean Air Congress, Düsseldorf, pp. A36-A39. Published by VDI.
(10) THIELE,V., E.KOCH & B.PRINZ (1985). Bestimmung von Geruchsstoffimmissionen mit Probanden. VDI Berichte 561, 313-328.
(11) MEDROW,W. (1985). Immissionsermittlungen mit Hilfe von Begehungen. VDI Berichte 561, 329-342.
(12) DE BREE,F.B.H. & H.HARSSEMA (1989). Application of sniffing teams within odour pollution research. In: L.J.Brasser & W.C.Mulder (Eds.), Man and his ecosystem; Proceedings of the 8th World Clean Air Congress, 105-110. Elsevier, Amsterdam.
(13) PERRIN,M.L., M.JEZEQUEL, J.L.DELPEUCH & R. NADAL (1989). Etude de la gêne provoqué par des odeurs d'origine industrielle. In: L.J. Brasser & W.C. Mulder (Eds.), Man and his ecosystem; Proceedings of the 8th Clean Air Congress, 111-116. Elsevier, Amsterdam.
(14) WINNEKE,G. & M.PADUCH (1989), Measurement and evaluation of odours in air quality control. In: L.J.Brasser & W.C. Mulder (Eds.), Man and his ecosystem; Proceedings of the 8th Clean Air Congress 1989. Elsevier, Amsterdam.
(15) KÖSTER,E.P., P.H.PUNTER, K.D.MAIWALD, J.BLAAUWBROEK & J.SCHAEFER (1985). Direct scaling of odour annoyance by population panels. VDI Berichte 561, 299-312.
(16) VDI 3940 (1989). Bestimmung der Geruchsstoffimmission durch Begehungen (Draft, August 1989, in preparation).
(17) KOCH,E. & G.SCHüTZ (1988). Gerüche. In: VDI-Kommission Reinhaltung der Luft (Publ.), Stadtklima und Luftreinhaltung, ein wissenschaftliches Handbuch für die Praxis der Umweltplanung, 386-403. Springer Verlag, Berlin, 1988.
(18) HEERES,P. & HARSSEMA,H. (1990). Progress of standardization of olfactometers in The Netherlands. To appear in Staub Reinh. Luft, May 1990.
(19) PRINZ,B., OTTERBECK,K. & KOCH,E. (1990). Überlegungen in NRW zur Bewertung von Geruchsstoffimmissionen. To be published by VDI.

ACHIEVEMENTS OF THE ODOUR GROUP

IR. J.H. VOORBURG
Institute of Agricultural Engineering (IMAG)
P.O. Box 43
6700 AA Wageningen, the Netherlands

In 1983, in the framework of the COST 681 programme, an "ad hoc" expert group was formed in order to evaluate the problems of "odour measurement and control" and to provide synoptic documents on the odour control and on the problems of the standardization of organoleptic measurements for assessing the odour emission or immission.

The expert group started with an inventory of existing guidelines for olfactometric measurements. The results of the inventory were presented in 1985 at Silsoe during a workshop on "Odour prevention and control of organic sludge and livestock farming". During this workshop the participating experts on odour measurement produced the "Recommendations on olfactometric measurements". In the introduction to the "Recommendations" it was said that the "Recommendations" are minimum conditions for those who wish to have consistent results and measurements which can be compared with odour measurements in other laboratories and other countries. Also laboratories beginning with odour measurements are advised to start at least at this level.

The experiences with and comments on the "Recommendations" were discussed during a workshop held in Zürich in 1988.

This led to the: "Improved Recommendations on Olfactometric Measurements" (1989). During the workshop attention was also paid to intensity scaling of odours.

A proposal for harmonization of odour intensity scaling will be discussed here. This proposal is prepared by a special sub-group.

Another special sub-group was formed to produce a proposal for field measurements of odours with panels. The results of this sub-group will also be presented here.

ACHIEVEMENTS OF THE ODOUR GROUP

Those who have participated in a ring-test with olfactometers and have seen the enormous variation in results between laboratories, will realize that the production and publication of recommendations does not give any guarantee for results which can be compared with odour measurements by other laboratories and in other countries.

Therefore, the recommendations should be considered a first step and a basis for communication and cooperation between institutes.

As such, the existing recommendations have proved to be useful. One of the members of the odour group even presented them at the 80th Annual Meeting of the APCA in New York.

This does not mean that the recommendations cannot be improved further. An example of the weak points in the recommendations is the expression of the odour concentration in odour units per cubic meter (OU/m^3). In fact this is nonsense because it is a number of dilutions that does not have any dimension.

With this example I will not plead for continuation of the work of the odour group on this part of activities. I think further improvement of the recommendations into guidelines is possible.

However, this can only be realized in combination with ring-tests.
If such a complicated effort should be realized at EC level it can better
be organized by a specialized laboratory.
This existing cooperation between experts from a number of countries in
the EC can be a basis for such an activity.
For bilateral cooperation between institutes the recommendations have
proved to be very useful. We have learned this from the excellent cooperation between IGAP at Hurley, NIAE at Silsoe and IMAG at Wageningen.

PROPOSAL FOR FURTHER ACTIVITIES
Important aspects of odour measurement are the sampling of odorous
air and the measuring of the volume of polluted air. From the beginning of
its activities, the odour group has been aware of the the the fact that
comparable problems are met in measuring other volatile emissions.
So, the odour group proposed to combine odour measurements from animal
manure or livestock production systems with the measurements of ammonia
emissions. Already during the workshop in Silsoe, Ryden described methods
for measuring ammonia emission from fields.
In 1987 a workshop was organized at Uppsala, devoted to volatile emissions. During this workshop most attention was paid to ammonia evaporation.
Meanwhile it is more and more realized that ammonia lost from livestock
operations contributes to the problems connected with "acid precipitation". Moreover, ammonia deposition also contributes to an overferti-
lization with nitrogen of natural vegetations and to the leaching of
nitrates.
So, more and more attention is paid to measuring ammonia emissions and the
development of systems that reduce ammonia losses.
As ammonia is an air pollutant, its dispersion is not limited to national
boundaries. So international cooperation in research on control of ammonia
emission should be considered.
Therefore, the odour group proposes to continue its activities under the
name of "volatile emissions"
This means it is proposed to reduce the attention for odour measurement
and to give first priority to development and harmonization of ammonia
measurements.
Important problems that deserve further attention are:
1. Measurement of open systems such as manure storage facilities or
   livestock houses with natural ventilation.
2. Measurement of fluctuating emissions.
3. Harmonization of measurement of volume of polluted air, sampling
   techniques and analyses.
I hope the discussion on this short introduction will show whether there
is agreement on this proposal and what the comments of the participants of
this workshop on the proposed priorities are.

REFERENCES
Hangartner, M., J. Hartung, M. Paduch, B.F. Pain and J.H. Voorburg. 1989.
Improved Recommendations on Olfactometric Measurements. Environmental
Technology Letters. Vol 10, pp 231-236.

CONCLUSIONS AND RECOMMENDATIONS

V C Nielsen[1] and B F Pain[2]

[1]ADAS Unit, Wrest Park, Silsoe, Bedford MK45 4HS, UK

[2]IGER, Hurley, Maidenhead SL6 5LR, UK

SESSION I

In this session the papers set out the problems which have to be solved. They emphasised the urgent need to reduce both ammonia and odour emissions from livestock production. The main reasons for concern are: (i) the severe damage which is caused by ammonia emissions to the ecology of natural plant habitats and forests, (ii) the increasing number of odour complaints by the public which is unacceptable and (iii) the financial losses which occur by the removal of nitrogen by ammonia volatilisation which then has to be replaced by nitrogen fertilisers.

SESSION II

Attempts to quantify the emission of ammonia from buildings and stores indicated that there were difficulties in obtaining accurate, representative measurements from all housing systems and large manure or slurry stores. Methods for measuring controlled environment buildings were well defined, problems arose when measuring naturally ventilated buildings.

There was great variability in the unit of measurement used ranging, from kg of ammonia per animal place a year to kg of liveweight per year. There was no agreement on the method of measuring ammonia emissions, those used ranged from spot checks to continuous monitoring. The emission rate is not constant but varies with time, so it is essential to have continuous monitoring for accurate results.

Papers discussed the range of possibilities which could be used to reduce emissions in buildings. It was generally agreed flushing was most effective for cattle and pig buildings and rapid air drying for poultry. Treatment by aeration following flushing was required to convert the ammonium salts to nitrates. Other treatments which look promising were acidification and manipulation of the nitrogen content of diets for pigs and poultry to reduce nitrogen excretion.

Recommendations

Future research and agreement is needed to determine the most reliable method of measuring ammonia emissions from naturally ventilated buildings and large manure stores.

Agreement is required on the use of standard units in ammonia measurements.

More research is needed to obtain reductions in ammonia and odour emissions from a variety of housing systems for the various types of livestock. There is a need to combine control methods, since it is unlikely that a single treatment system will achieve a significant reduction alone and at a cost to the livestock enterprise which can be sustained.

SESSION III

Research on the use of bioscrubbers indicated that successful systems are being developed which control ammonia and odour emissions. Self cleaning biological treatment is difficult to achieve and sustain. A very high level of technical input is required at present, and systems are not yet user friendly.

Biofilters have proved successful but again require a high level of technical skill to maintain optimum conditions. Present successful systems are too expensive for use on farms.

Recommendations

Further development is necessary to achieve cheaper reliable systems.

Cheap and effective monitoring systems are needed to indicate to the operator the state of the plant and to determine if there is a need to alter the balance of nutrients.

SESSION IV

The emissions of ammonia and odour during and following the spreading of manures and slurries accounts for between 50 - 70 per cent of all emissions. There are a large number of methods of measuring emissions in use, ranging from small enclosures to micrometeorological techniques.

There are now available a range of techniques for reducing ammonia and odour emissions during and after application to land including a wide range of machinery. At present, little attention has been made to determining whether the retained ammonia remains within the soil-plant system or is lost by leaching or by gaseous emission into the atmosphere.

Emission of ammonia and odour is influenced by a wide range of factors relating to environmental conditions, waste management and waste composition.

Recommendations

There is a need to compare the various methods of measuring ammonia emissions during and following manure and slurry application and to make recommendations on the most reliable and accurate methods of measurement. Also, to obtain agreement on standardising techniques to enable comparisons of measurements.

There is a need to set out recommendations on suitable techniques which evaluate the performance and reduction of ammonia and odour for manure and slurry spreading machinery. This could then enable farmers to compare similar machines and to evaluate the most effective one for his particular soil and climate. Also to compare the costs of different techniques, for example shallow injection compared with dilution by irrigation.

SESSION V

There are great difficulties in measuring ammonia losses from grazed pasture. It is necessary to measure losses over the entire grazing period when ammonia losses appear to be closely related to nitrogen inputs. At present research is concentrating on the effects of less nitrogen input systems relying on balanced clover grass swards to maintain production rather than traditional high mineral fertiliser systems.

Recommendations

Further research on natural systems needs to be carried out to find suitable systems for different climatic conditions. Those suitable for Northern Europe are unlikely to be effective in dry Southern European Countries.

SESSION VI

This session was concerned with odour emissions. Agreement on recommended methods of olfactometric measurements have been achieved and include odour concentration. It was not possible to reach agreement on odour intensity which is an important aspect of the characteristics of an odour. A method of determining the offensiveness of an odour and its nuisance using a panel of individuals down wind of the source was agreed. This method is cheap and effective and could be of value in situations where a decision is required without undue cost.

Recommendations

There is a need to determine rates of odour emission for a range of livestock housing systems and for manure and slurry storage. This would be of practical value to livestock farmers in determining the most appropriate housing system for their particular situation. At present livestock housing is designed to optimise the environment in a building which provides heat, light and animal welfare consistent with low inputs of energy and feed to achieve optimum profits. There is no rating given for the impact of such housing systems on the environment. Ammonia and odour emission concentrations are not available. Until these parameters are readily available, there will be no practical way of reducing these emissions from livestock production.

Further attempts should be made to produce recommendations for measuring odour intensity. More attention should be paid to the presentation of results from olfactometric measurements. Currently, the odour threshold value is frequently expressed as odour units per m^3 air, although it is really a dimensionless value i.e. a number of dilutions.

LIST OF PARTICIPANTS

A. AMBERGER
University of Munchen
Institute of Plant Nutrition
D-8050 Freising-Weihenstaphan

B. BERGLUND
University of Stockholm
Department of Psychology
S-106 92 Stockholm

H. BOMANS
K.U.Leuven
Fakulteit Der Landbouw
Wetenschappen
Laboratorium Bodemvrucht-
baaheid & Bodembiologie
Kardinaal Mercierlaan 92
B-3030 Leuven

M.A. BRUINS
IMAG
P.O.Box 43
NL-6700 AA Wageningen

O. CARTON
Agriculture and Food
Development Authority
Johnstown Castle Research
Centre
IRL- Wexford

W.J. CHARDON
Institute for Soil Fertility
Postbox 30003
NL-9750 RA Haren

C.R. CLARKSON
Institute for Grassland
and Animal Production
Hurley Maidenhead
UK-SL6 5LR Berkshire

M. DE BODE
IMAG
P.O.Box 43
NL-6700 AA Wageningen

P. DEGOBERT
Institut Francais du Petrole
1&4 Av.de Bois-Preau BP 311
F-92506 Rueil Malmaison Cedex

H. DOHLER
University of Bayreuth
Department of Agroecology
Postbox 101251
D-8580 Bayreuth

M.R. EVANS
ICI Brixham Laboratory
Freshwater Quay
UK- Brixham, Devon

M. FERM
Swedish Environmental
Research Institute
P.O.Box 47086
S-402 58 Gothenburg

J.H. HALL
Water Research Centre
Medmenham Laboratory
POBox 16 Marlow
UK-SL7 2HD Bucks

H. HARSSEMA
Department of Air Pollution
University of Wageningen
P.O.Box 8129
NL-6700 EV Wageningen

J. HARTUNG
Institut fur Tierhygiene
Bunteweg 17P
D-3000 Hannover 71

J. HOBSON
Water Research centre
Frankland Road POBox 85
UK-SN5 8YR Blagrove, Swidon

J. JANSEN
FAL
Institut fur Biosystemtechnik
Bundesallee 50
D-3300 Braunschweig

S.C. JARVIS
Institute for Grassland
and Animal Production
Hurley, Maidenhead
UK-SL6 5LR Berkshire

J.V. KLARENBEEK
IMAG
P.O.Box 43
NL-6700 AA Wageningen

A. KLASINK
LUFA
Mars le Tour-Str 4
D-2900 Oldenburg

E. KOCH
LIS
Wallneyer Strabe 6
D-4300 Essen 1

E.P. KOSTER
Vakgroep Psychische
Functieleer
Universiteit Utrecht
Sorbonnelaan 16
NL- Utrecht

K.H. KRAUSE
FAL
Institut fur Biosystemtechnik
Bundesallee 50
D-3300 Braunschweig

P. L'HERMITE
Commission of the
European Communities
Environment and Waste
Recycling
Rue de la Loi 200
B-1049 Brussels

G. LUNDIN
Swedish Institute of
Agricultural Engineering
Box 7033
S-75007 Uppsala

H. MANNEBECK
Institute of Agricultural
Engineering
University of Kiel
Olshausenstr. 40
D-2300 Kiel

G.J. MONTENY
DLO
P.O.Box 59
NL-6700 AB

V.C. NIELSEN
ADAS Unit
Wrest Park, Silsoe
UK-MK454HS Bedford

J. OOSTHOEK
IMAG
P.O.Box 43
NL-6700 AA Wageningen

M. PADUCH
Verein Deutscher Ingenieure
Graf-Recke Strasse 84
D-4000 Dusseldorf

B.F. PAIN
Institute for Grassland
and Animal Production
Hurley, Maidenhead
UK-SL6 5LR Berkshire

C. PEARSON
ADAS
79/81 Basingstoke Road
UK-RG2 OEF Reading

M.L. PERRIN
Inst. de Protection
et Surete Nucleaire
Dpt de protection technique
CEN/FAR BP 6
F-92265 Fontenoy-aux-Roses

V.R. PHILLIPS
Institute of Engineering
Research
Wrest Park
UK-MK454HS Silsoe, Bedford

J. ROELOFS
University of Nijmagen
Vakgroep Aguastische
Ecologie en Biologie
Toernooiveld 1
NL-6524 ED Nijmegen

S. SCHIRZ
KTBL
Landeschlayer Str. 18
D-6100 Darmstadt

C. SCHMIDT VAN RIEL
IMAG
Post Box 43
NL-6700 AA Wageningen

R. SCHOLTENS
IMAG
P.O.Box 43
B-6700 Wageningen

S.G. SOMMER
Askov Experimental Station
Askov, Vejenvej 55
DK-6600 Vejen

L. SVENSSON
Swedish Institute of
Agricultural Engineering
Box 7033
S-75007 Uppsala

L. VALLI
Centro Ricerche
Produzione Animale
Via Crispi 3
I-42100 Reggio Emilio

N. VERTREGT
CABO
P.O.Box 14 - Bornsesteeg 65
B-6700 AA Wageningen

K. VLASSAK
K.U.Leuven
Fakulteit der Landbouw
Wetenschappen
Laboratorium Bodemvrucht-
baaheid & Bodembiologie
Kardinal Mercierlaan 92
B-3030 Leuven

J.H. VOORBURG
IMAG
P.O.Box 43
NL-6700 AA Wageningen

INDEX OF AUTHORS